From Mechanism to Organism
– Enlivening the Study of Human Biology –

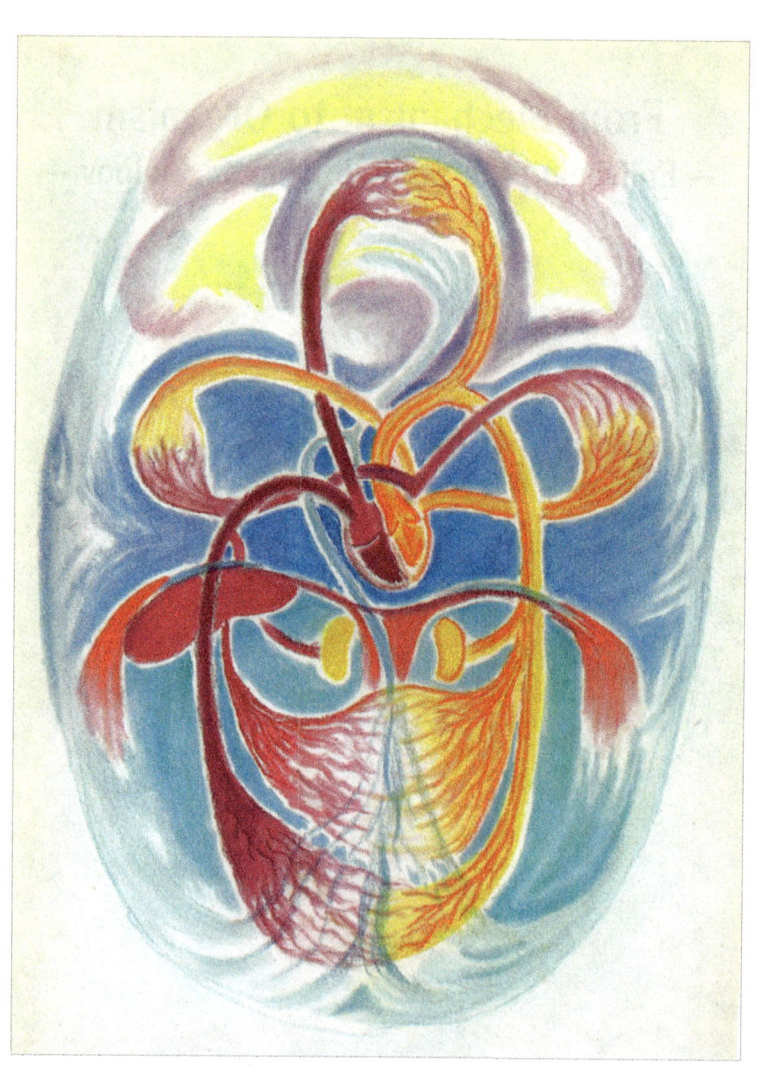

From Mechanism to Organism
— Enlivening the Study of Human Biology —

Michael Holdrege

Printed with support from the Waldorf Curriculum Fund

Published by
Waldorf Publications at the
Research Institute for Waldorf Education
351 Fairview Avenue
Suite 625
Hudson, NY 12534

Title: *From Mechanism to Organism*
  Enlivening the Study of Human Biology
Author: Michael Holdrege

Drawings: Michael Holdrege
Proofreader: Alice Brown
Layout: Ann Erwin
Cover image: Adobe Stock

© 2022 by Waldorf Publications
ISBN #978-1-943582-66-2

Table of Contents

Preface 7

Introduction: Developing Sound and Independent Judgment 9

PART I. Ninth Grade Human Biology

    Pedagogical Perspectives 16
    Getting Started 19
    The Eye and Seeing 22
    The Ear and Hearing 36
    The Sense of Balance 47
    The Sense of Movement 53
    The Larynx and Speech 59
    The Human Skeleton
        Overview 64
        The Limb Pole 67
        The Head Pole 72
        The Torso 75
        Uprightness 78
    Joints 83
    Muscular System
        Leverage 88
        Different Levels of Muscular Activity 92
    Bibliography – Part I 99

INTERMEZZO: Waldorf Biology in a Reductionist Setting 103

PART II. Tenth Grade Human Biology

    Process-based Understanding 112
    Getting Started 114
    The Cardiovascular System (CVS) 116
        Circulation – the Flow 118
        Arteries and Veins 121

Capillaries	124
The Heart	126
Peripheral Circulation	136
Microcirculation and Flowback	140
Sensing and Harmonizing	143
Polarities and Rhythms	144
Inner and Outer	147
Warmth	149
Coronary Circulation	152
Blood Pressure	154
Human Blood	156
The Immune System	
The Non-specific (Innate) Immune System	166
The Specific (Adaptive) Immune System	168
HIV/AIDS	174
The Respiratory System	179
The Brain	186
The Digestive System	
The Mouth and Esophagus	203
The Stomach	206
The Small Intestine	208
The Large Intestine (Colon)	212
The Microbiome	213
The Liver	222
The Kidneys	228
Looking Back	235
Bibliography – Part II	242
Index	248
Acknowledgments	252

Preface

In the many years that I have been involved in Waldorf teacher education (almost forty by now), one thought has continued to echo in my mind: New teachers need *examples*. I heard it first during my teacher training from the well-known Waldorf teacher, adult educator and author, Christof Lindenberg.[1]

What he meant by this was not prescriptive, it was practical. Lindenberg understood that even when teachers have a solid grasp of the subject matter that is suitable for children at a certain age—complemented by an appreciation of how different subject areas meet their needs in different ways—knowing how to embody such insights into concrete teaching material is no small thing. For that reason, it is very helpful for newer teachers (and experienced ones too), to hear how someone has actually done it, has "incarnated" the general idea of what to teach into a particular approach with a particular emphasis. Lindenberg was a history teacher and he gave many *examples* of how one could teach this or that aspect of history. He was not saying that we had to teach about Mao and "the long march," or that we had to do it in the way he did it, but through his *examples* he stimulated our imagination to see how things could be taught in ways we had never thought of before.

Years later, when I reflected back on those experiences in teacher training, I realized how important what Lindenberg had said had been for me when it came to making the great leap from student to teacher.[2] Although the essentials of Waldorf education were the foundation, the *examples* were key in learning to move from large ideas to concrete applications. I realized, for *example*, how central for my zoology teaching it had been to experience Ernst-Michael Kranich climbing up on a table and on all fours roaring like a lion until the whole building shook! Even though I have never been talented enough to match such a roar, it did loosen my imagination enough that I began trying some of my own (silly) animal imitations with the students, such as hopping with the class down the hallway as if we were frogs. Dr. Kranich was huge for me in so many ways

---

[1] Among his numerous publications, Lindenberg is also the author of *Teaching History: Suggested Themes of the Curriculum in Waldorf Schools.* Hudson, NY: Waldorf Publications, 1989.
[2] Back in those days I was also a history teacher.

Preface

through his *examples* of how one can approach the study of plants, of animals, of the human being. Much the same can be said about reading the many *examples* given in these subject areas by Kranich's mentor and colleague Fritz Julius. Life changing! *Examples* given to me by Wolfgang Schad were decisive in helping me prepare for my first 9th grade biology block and provided a perspective that informed my teaching for years to come, even when the specifics morphed from year to year depending on the group of students in front of me. Once I had my feet on the ground in the classroom, my colleague Felix Bauer provided many more enlivening examples that furthered my teaching.

I have also observed in the context of mentoring and teacher training that *examples* are what people very often ask for. What do you do here? How do you approach this and that? Again, it is not about a set approach, about the "only" or the "best" way, but about stimulating teachers to try new things, to hop out of their ruts (we all have several) and find new avenues into this or that topic. I see this kind of stimulus—when effective—as planting a seed that can unfold with each teacher's own signature and flavor over time. (And the ideas that don't enliven are easily weeded out!)

So, in that spirit, I decided to write down some of the ways that I have approached many of the topics in the Waldorf high school biology curriculum. I have done this not because my *examples* are brilliant in any way or manner—so naïve am I not—but in the spirit of stimulation for those who have recently taken on the task of teaching these blocks and are looking for ideas to help them get started on this path, as well as for others who have been teaching for a while and would like to bring some new flavors here or there into the topics they teach. It is from that perspective that I pass on to the reader some *examples* of what might be done in various biology blocks, grades nine through twelve, in a Waldorf high school.

– MH

INTRODUCTION:
Developing Sound and Independent Judgment

As anyone who has ever taught high school students knows, there are significant changes in their interests and capacities that appear as they move through the grades 9 to 12. In Waldorf education we try to shape our teaching and choose our subject matter so that it resonates with the age-group we are teaching. This stands in contrast to much of modern education, which does not emphasize developmental phases and mainly teaches test-pertinent topics and test-relevant techniques. A central question for the Waldorf teacher is, by contrast: What is the 9th grade "state of mind" and how do I recognize and meet it in a way that "fits" the students where they are, that corresponds to what they are able to internalize and digest and is therefore supportive of their development?

As is well known, adolescents entering high school are intensely engaged in finding a new relationship to the world around them. They are immersed in a process that Harvard professor Robert Kegan calls "meaning-making": They are struggling to make sense of their life experiences in a new way, as well as find their own place within them (Kegan 1982). It is clear that young people today do not simply want to absorb values and an understanding of the world that comes from someone else or from tradition. They want to make their own judgments, based on their own experiences, and they want to give their lives direction on that basis (aka self-determination). The desire to form their own individual judgments and draw their own conclusions is a mark-stone of this age group and appears in conjunction with what Piaget calls the birth of "hypothetical/deductive thinking" (formal operations). This capacity enables them to envision a world that is different from (better than) the one they find themselves in. They can imagine hypothetical situations and deduce what would follow from them if they were in fact "real" (Piaget & Inhelder 2000).

This capacity—although central for the birth of independent judgment—is anything but balanced at the outset. As we all know, judgments—outspoken or not—come hard and fast from young people in this age group. This new capacity must be practiced and tested, because students are now able to think about things in terms of what "could be" rather than how they actually are. Their views of things can be very quick, critical and egocentric at times, but they "belong" to

Introduction

them! They are born out of their newfound capacity to judge for themselves, and they are carried by feelings they experience as being their very own.

Another key aspect of the new capacity to see the world as it "could be," is that it provides a very real basis for idealism. Do I just accept the status quo, the imperfect world created by older generations, or do I join the battle for a new future that treats people more humanely and provides opportunities for all? Should we strive to find a new relationship to nature, or is she merely something to be exploited, and so on? This capacity to create ideas and intentions that transcend the imperfect realities of the world as we know it can, of course, lose the ground under its feet and fall out of balance if carried too far.

One of the central tasks for Waldorf teachers working with this age group is to help students gradually develop—through the four years of their high school education—a strong, differentiated and individualized capacity for healthy judgments and reality-based ideals. In other words, we work with the student's capacity to encounter the world in its ever-changing, endlessly diverse manifestations and help them to develop concepts that accurately and sufficiently illuminate whatever it is they are faced with or engaged in.

What is generally termed "phenomenology" in Waldorf education refers to a process where students are presented with concrete phenomena that—before they have been understood—appear to them as riddles. If such "riddles" are presented to them in an interesting way, they awaken in the students a desire to understand, to pursue concepts that shed light on the mysteries before them. This is sometimes referred to as "learning from the inside out," in contrast to extrinsically motivated learning that only takes in information as a means for achieving some other end (such as getting a good grade).

Real learning is, however, not usually just a "one-moment" occurrence. During the process of mature, healthy judgment formation, concept and percept interact in a kind of dialogue. When I meet some thing, situation, person or topic, I try to perceive clearly what I have before me, to feel how it speaks to me, and to relate it to other experiences I have had (or information available to me). After pondering such factors and their relationships, I then bring forth a concept that helps me grasp more deeply the issue at hand. Healthy judgment isn't finished yet, however. It "listens" further: Does the percept really accept this concept, or is there a dissonance somewhere? I bring forth another concept—

does it fit better? (and so on)—until I sense, YES, now my thinking has indeed brought forth the concept (or network of concepts) that reveals something essential about this riddle before me—they belong together!

Yes, many will say, this sounds great, but how does one cover all the necessary subject matter for a particular grade level when one works so thoroughly? This presents for conventional biology teachers in the U.S. considerable challenges. If we look, for example, at a widely used biology textbook for grades 9 and 10, Miller & Levine's, *Biology* (2010), we find 1034 pages of subject matter (that weighs 6.9 pounds and costs $135) to be covered in two school years. The problem this presents is one that many thoughtful educators have reflected upon over the years. One in particular, Martin Wagenschein, spent decades illuminating how the principle of "less is more" often makes sense in modern society, where people are inundated with information from all directions on a daily basis. Wagenschein was a professor of education at the University of Tübingen and well-known in German-speaking education circles. He saw the detrimental effects of rote learning and theory-based teaching that are so prevalent in our time. In contrast to this, he developed an experience-based approach to science education, one which is focused on the exploration of concrete phenomena in a way that helps learning become a process of inquiry rather than just the absorption of set facts and theories. Below are a few excerpts from Wagenschein's writings that highlight a way of teaching that supports a thorough judgment formation process on a phenomenological basis.

> *Let us begin by looking at what we have to steer away from if school is not to suffocate from the sheer mass of content and then perish as a kind of subject-matter processing plant. The older and more established a subject is, the stricter we tend to plan the learning steps. In these approaches, the essential thing seems to be: Every single detail serves as a small stepping stone, leading the learner to something more complex and difficult, which he or she cannot yet grasp. The reasons for working in this way are obvious. One thing builds on another, either logically or chronologically. There has to be order. If you leave something out, you will have to pay for it later. Every detail can be important, even if you don't recognize how at the moment. These reasons are "logical," but that is all they are. They are not pedagogical. They only see the totality of the*

Introduction

subject matter and overlook the student. The student is seen as a small adult, quantitatively limited in their ability to grasp things.

But to be a teacher means: to have a feeling for the process of human development, for what the human being is growing toward, for the awakening spirit. And to be a subject teacher means to know, in addition, both the subject's being and its becoming within the learner.

… An inherent temptation is completeness, which leads to haste and a lack of thoroughness. And an impressive heap of gravel is thus built up. Education is not a process of just adding. … So what we need are selection criteria. We need to confine ourselves to the essential. … We recommend the courage to leave gaps, which means the courage to be thorough and to dwell intensively on selected topics. So instead of evenly and superficially walking through the catalog of knowledge, step-by-step, we exert the right —or fulfill the duty—to really settle in somewhere, to dig in, to grow roots and take root. The particular aspect we delve into is not a stage in a process, but a mirror of the whole.

Why? The relation the particular has to the whole is not that of a part, step, or preamble; it is a center of gravity. It may not be only one, but it carries the whole in it. This single aspect is not an element in a process of accumulation, rather, it carries and illuminates. It is not a stage in a progression, but it works like a spotlight. It affects things that are distant yet related through resonance. This is what the concept of the exemplary means. … (1982)[3]

Wagenschein's exemplary approach is, in my experience, one of the keys to effective teaching in a Waldorf high school. It enables the teacher to work with the subject matter in such a way that judgment formation can be practiced and schooled at many levels over many years. It helps eliminate the pedagogically unfruitful (even damaging) haste that many teachers fall into in order to "cover all the material" they are "supposed to" in a particular subject area.

---

[3] For several additional essays by Wagenschein on exemplary teaching, go to the Nature Institute website: natureinstitute.org

Viewed from Wagenschein's exemplary perspective, the many subjects that the students engage through their high school education can be seen as working together—like the various instruments in an orchestra—to help bring about a multi-faceted capacity for judgment formation, one that each subject by itself would never be able to nurture and bring about. One need only think of the value of the arts in developing observational sensitivity and imagination—which are central prerequisites for sound judgment—in order to realize the wonderful possibilities that the chorus of Waldorf high school subjects bears within it.

PART I

Ninth Grade Human Biology

Part I. 9th Grade

Pedagogical Perspectives

What topics from the field of biology are well-suited for helping 9th grade students develop their capacity for independent judgment? Decades of experience in Waldorf schools around the world confirmed that the sense organs and the skeletal/muscular system are an excellent choice for this task.[1] Why? As always, the background question for the Waldorf teacher is: What relationship do the students have to the world around them and to themselves in their current phase of development? What are the central riddles that they face at this age? Two deep-seated changes in their relationship to the world that inform their sense of well-being as 9th graders can be roughly characterized as follows:

1) In great contrast to when they were younger, they experience their physical bodies as uncoordinated and out of harmony with their surroundings.

2) They experience a new distance[2]—a lack of spontaneous connectedness—to much of what they experience in the sense world around them.

The birth of hypothetical/deductive reasoning referred to above can be seen in conjunction with how they now inhabit their bodies and with their changing experience of the outer world. Rudolf Steiner points to one decisive aspect of this:

> *If you observe children of over 12 years old you will see from the way they step how they are trying to find their balance, how they are inwardly adapting themselves to the leverage and balance, to the mechanical nature of the skeletal system. ... And only now that they have taken hold of that remotest part of their humanity, the bone-system, does their adaptation to the outer world become complete. Only now are they a true child of the world, only now must they live with*

---

1 These are the same themes taught by the class teacher in the 8th grade. Although the students may at first wonder why the same area is being considered again, it provides the high school teacher with an opportunity to show them that in high school they are dealing with "real" science. They should experience that a wealth of new insights is now available to them through the approach and expertise of their high school teacher (Richter 2016).

2 In Waldorf terminology, these are an indication of emerging antipathy forces.

the mechanics and dynamic of the world, only now do they experience what is called 'causality' in life. (1922)

Why do they, at this age, live more strongly in their sense of balance and in the mechanical/leverage forces of their skeletal system, which, as Rudolf Steiner says, causes them to experience the nature of causality at a new level? This has to do with the rapid growth of the limbs in adolescence, which causes them to fall out of the harmonious hopping, jumping, skipping movements of their middle-school years. In contrast to their earlier growth from the head downward (cephalo-caudal), they now begin to grow from the periphery inward (distal-proximal): At first the feet and hands grow larger, then the arms and legs, and lastly the rest of their bodies. They suddenly have leverage issues! Their limb bones have grown longer and heavier, but there has been no corresponding increase in muscle mass—which is delayed approximately one year—that they need to meet these issues. (Their load arm has increased in length relative to their lift arm, thus requiring more effort than they needed to lift the very same object when they were younger [Leber 1993].)

For this reason, their movements become very inharmonious. Until around the age of 16, they struggle to bring their own bodily movements into accordance with the strict causality of the mechanical and gravitational forces of the outer world. Longer limbs mean dealing with gravity at a whole new level. In conjunction with these deep-seated, only semi-conscious experiences, a sense for the unambiguous nature of causality also begins to dawn in their thinking.

For these reasons, the examination of the human skeletal/muscular system meets deeply-rooted, but only dimly-conscious, questions that live in them at this age. Using the kind of causal thinking that explains clearly the functioning of the skeletal and muscular systems can be very gratifying for them, since it sheds light on the very real bodily challenges that they are going through at this time in their lives and helps them develop confidence in their thinking capacity.

At another level, it is through the senses that the students receive a constant, never-ending array of impressions from the world around them. These impressions are the basis for their conscious life on earth and provide the context for all their activities. They come to them from without and must be met from within through their own formative, organizing, thinking activity. As Owen Barfield (1988, p. 26) points out, the pure sensations that come through

Part I. 9th Grade

the senses *must be combined and constructed by the percipient into recognizable and nameable objects we call "things."* Barfield gives this constructive activity the name "figuration." He illustrates it with a humorous example from Lewis Carroll's old jingle, *Sylvie and Bruno*, that has a constant refrain "he thought he saw," followed by "he found it was":

> *He thought he saw a Banker's clerk,*
> *Descending from a bus,*
> *He looked again and found it was*
> *A hippopotamus…*

To focus on this internal, but constitutive aspect of their perceptual experience, however, would not fit the 9th grader's stage of inner development (although we will touch upon it briefly in the context of the eyes and seeing). On the other hand, addressing the organs that mediate the conscious experience of the world they have awakened to is very appropriate. The basis for this lies in the fact that teenagers find themselves in a considerably more distanced and sober relationship to many aspects of their surroundings than they did as preteens.

Moreover, the human sense organs lend themselves to straightforward causal explanations of their structures and functioning. They can be shown to exemplify clear-cut linear causality. As has been said, it is just this form of thinking that we find at the forefront of the 9th grade mind. They have a predisposition to think in this way, and finding that those thought-forms shed vivid light on a key aspect of their relationship to the world (sense perception) helps them build confidence in this capacity. From this perspective, the main objective of a 9th grade biology block is not to simply fill the students with lots of knowledge about our sensory organization, but to fill them with the experience (and the confidence that comes from it) that their thinking is capable of shedding light on many complex aspects of life and hence an essential "tool" to be exercised with joy and great benefit.

Getting Started – A Possible Day One

As most block teachers will have experienced, the first day is pivotal. Getting off to a good start can create momentum that makes all the difference in awakening student anticipation for what will come in the weeks ahead. This is particularly true when—as is often the case in the 9th grade biology block—this is the first time that the class and the teacher have ever worked together. A lot depends on the specific group of children, of course, but I have usually found that the students look forward to their first life science block in high school. For that reason, I always emphasize that science in high school will be at a new level, that even though they may have covered some of the same topics in grade school, these will be dealt with now in much more depth and rigor than ever before! (This kind of approach is not meant to be daunting, but to meet the students' more or less conscious expectation that their high school experience will be challenging and filled with cutting edge content. If they don't get this impression, they will be disappointed and at some level wonder if their Waldorf high school experience will really prepare them for "the real world" that is out there waiting for them.)

One possible way to begin this first meeting is to ask them what they think biologists actually study. After some discussion, we begin a list of the various realms—sub-disciplines—of biology that they have heard of. These can be put on the board, resulting in an inventory that goes something like this: botany, zoology, genetics, evolution, anatomy, physiology, microbiology, ecology, ornithology, embryology, etc. Once the list is compiled and, if necessary, supplemented, the students are usually excited to hear that all these areas—some of them with daunting names—will be touched upon in the biology blocks over the next four years. I then turn to the list and write next to each sub-discipline the grade level/block in which this topic will appear. This makes an impression and lets them know that there will be lots of real science ahead of them in the years to come—and it's starting right now!

After an easygoing warmup such as the above, I have found it worthwhile to prepare the ground for the 9th and 10th grade biology blocks by asking the students how much they actually know—out of their own direct experience—about various organs and organ systems of the human body. After fielding numerous contributions, it is possible to draw out the distinction between what they know first-hand and what they been taught or told about by others (the experts). After some discussion the distinction can be made between organs/parts of the body that:

- a) we know about first hand (eyes, arms, etc.),
- b) we sense to some degree (lungs-breathing, stomach-hunger, heart-beat),
- c) we know nothing at all about from direct experience (liver, spleen, etc.)—although we might experience them—unpleasantly—when something is wrong with them!

With respect to b) and c), what we know we have only learned through the research of others. A further question can be: Which of these organs do we have conscious control over and which operate on their own, independent of our intentions. (There is also a category of organs between the two extremes that we can influence indirectly; for example, we can increase our heart rate by running laps, etc.)

Such discussion can lead to a rough and ready distinction between "outer organs," that we know about directly and can influence consciously, and "inner organs," that are largely beyond our influence and not known to us through direct perception.

One more question worth asking in this context is: What roles do these two categories of organs play in our lives? Is one group more important than the other?[3] What happens, for example, if an inner organ no longer functions? We die! We count on them to perform their respective activities day after day, year after year, until they don't and life is over. What happens if an outer organ doesn't work? We lose our experience of some aspect of the world around us, or we lose the capacity to engage that outer world through our organs of movement. (If they were all to shut down, we would live in total isolation—total darkness, total silence, no touch, nothingness!) Clearly, our quality of life, our experience of the

---

3   The idea of drawing the distinction between "outer" and "inner" organs as a starting point for the 9th grade block was given to me by Wolfgang Schad in the final days of teacher training, as I was desperately trying to figure out how to make the leap into my very first biology block. I have tried multiple variations of this approach since and still find it to be a great entry point for the 9th grade.

outer world and our relationship to it is shaped by the outer organs. The proper activity of the inner ones, on the other hand, is a matter of life or death!

After such considerations we can explain to the students that the 9th grade biology block will be dealing with areas of the body over which we have some conscious control and that are familiar to us (what we have called the "outer organs"), whereas in the 10th grade we will be exploring those aspects of our anatomy and physiology that relate to our "inner organs."

Part I. 9th Grade

The Eye and Seeing

From the perspectives just discussed, the human eye provides an excellent starting point for an exemplary investigation of the human senses. Not only can it be viewed easily from a variety of perspectives, but it also plays a major role in the lives of teenagers. Students often find it interesting to begin with direct observations of the eye, "face to face." The students pair up and observe carefully what they see, describing this to each other without using anatomical terms.

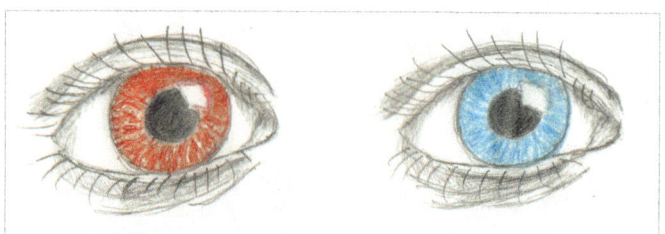

Fig. 1.1 Two human eyes, frontal view.

They notice the dark circle in the middle, the ring around it that (as they know) varies in color from person to person, and then the bright white area in the periphery that disappears under a layer of skin at its edges. Although familiar to everyone for as long as they can remember, they are surprised to discover how exact, focused observation can yield hitherto unnoticed details—and awaken new questions. After gathering such observations, we compare them with other parts of the human body and with nature in general. Do we find such "perfect circles" anywhere else? After some reflection, someone will usually point out that the round shapes we find are reminiscent of the earth, the sun, the full moon, and the planets (when seen with a telescope). These are, of course, the sources of light for life on earth. We consider the color variation in the ring around the dark center and ask if we find any comparable colors elsewhere in humans. And lastly, can we find whiteness comparable to the outer perimeter of the eye (the sclera) elsewhere on the surface of our body?

Such comparisons can lead to more general considerations such as: Whiteness comes from a strong reflection of the light, whereas black indicates either absence or complete absorption of light. It will usually occur to someone that the dark circle in the middle (the pupil) is actually an opening where light enters the eye. In other words, we are looking into a space that is filled with light—and yet it is pitch-dark! How is this possible? This question we take with us to ponder till the next day.

If I tell the students that even though most Caucasian children are born with blue eyes (and in four out of five cases they soon darken to brown) there is no such thing as blue pigment in the eye, an additional riddle arises. We reflect on whether anywhere else in nature shows such a striking blue. Before long, the deep blue sky on clear days comes into the discussion, and also the profound blue that bodies of water sometimes reveal. Do the sky and the blue water contain blue pigment? The students know, of course, that a bottle of blue Lake Michigan water is no longer blue after it leaves the lake—but why? This question, too, we take with us overnight.

After a review of the above observations the next morning, two things follow logically from the blackness of the pupil that opens into a light-filled space. First, light itself must be invisible, otherwise we would see it when it fills a space.[4] Second, since we see things when they reflect light, then the light in the inner eye must not be reflecting off of anything. Therefore something must be absorbing it completely. This awakens their curiosity—we hope—about what we will find when we explore the interior of the eye, since we can't see anything at all when we look into the dark pupil.

Before we do that, we have another question carried over from yesterday. Some people have blue irises—and most of us do at birth—and yet we have told the students that there is no blue pigment in the eye. We determined that it reminded us of the blue sky on a clear day, and of lakes and oceans on some days. Do they contain blue pigment? First of all, someone points out that the sky is black at night, and black for those who travel in outer space. We note that the color of the sky changes with outer conditions: deep gray on a cloudy day, blue on a clear one, black at night. Gradually we come to an insight first emphasized by Goethe in his *Theory of Color*: The semi-transparent, illuminated (during the daytime) medium—the atmosphere—that lies between us and the dark sky in

---

4 If students struggle with this idea, examples can be given of physics experiments in which light is shining through a completely dark room (from one tube into another) and yet no one can see if the light is on or not—unless there is something for it to reflect off. Or it can be pointed out that outer space is totally black, although filled with sunlight, as well as light from other sources.

space must be influencing the color. We then come to the conclusion that when we look through this illuminated medium at the darkness in the background, blue appears. The denser the medium, the paler the blue, the less dense—for example, high on a mountain-top on a clear day—the deeper the blue becomes, until it approaches violet. (If we kept going up into the sky it would eventually turn black.)[5]

Returning to the iris, we can now attempt to explain the blue color. Thinking this through we gradually come to the insight that we must be looking through a semi-transparent illuminated medium onto a dark background. And indeed, this is the case: The iris contains a dark pigment in its innermost layers and a semi-transparent, tissue externally. When we look at such an iris, we are looking through the external tissue (similar to the illuminated atmosphere) at a dark background (similar to outer space) and the color blue appears. This kind of color is not a pigment, not a substance, and for this reason it is commonly called a structural or atmospheric color.[6] Like the colors in the sky, it arises through the interplay of light and darkness.[7]

Once the students begin to realize how color can arise through the interaction of light and darkness, they are amazed to see how the human eye is actually an image of this: brightness in its periphery (sclera), darkness in its center (pupil), and between these poles a play of color ranging from atmospheric to earthly (from blues, to various gray tones, to browns). When we consider, in addition, the "sparkle" that can come from a human eye, the students realize what an amazing, light- and darkness-related confluence of qualities come together in this pair of light-receptive organs located front and center in the human head. Nowhere else do we find in the visible human being such a place of light, darkness, color and potential "sparkle," thereby mirroring both our inner life and the outer world.[8]

---

[5] It usually helps here to also describe the opposite phenomena: When a colorless light source (the sun) is viewed through the atmosphere, it becomes yellow; the denser this medium becomes, the more it moves toward orange and reddish hues.

[6] When a baby's eyes change from blue to brown, this is because brown pigment (melanin) enters the outer layers of the iris tissue and we see the brown pigment itself—which is no longer "atmospheric."

[7] See Goethe's *Theory of Color*, paragraphs 150–151.

From this more qualitative beginning, we now move to an exploration of the eye in search of clear causal explanations for what we find. Since the darkness of the eye's interior has created the question of what is actually in that space, it behooves us to create a large drawing (sagittal section)—that reveals the internal structure of the human eye.

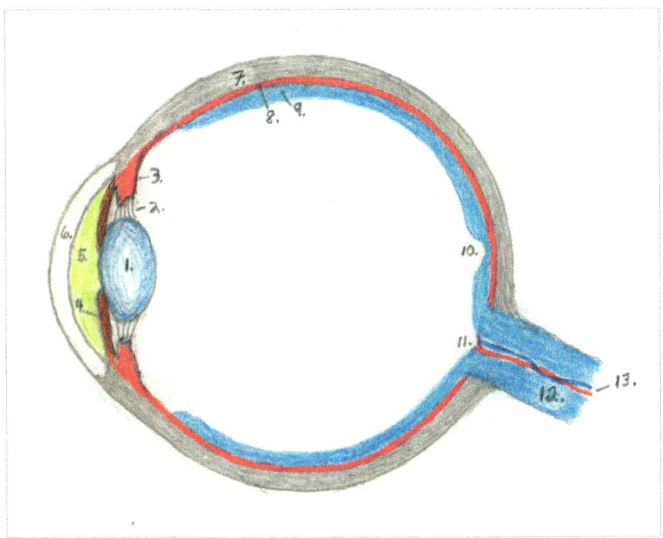

Fig. 1.2 Sagittal section of the human eye. 1. Lens. 2. Radial fibers. 3. Ciliary muscle. 4. Iris. 5. Anterior chamber. 6. Cornea. 7. Sclera. 8. Choroid. 9. Retina. 10. Fovea centralis. 11. Optic disc. 12. Optic nerve. 13. Central artery, central vein.

A preliminary characterization of the key components of the eye seen in our drawing provides a first rudimentary overview of the path of the light as it passes through the eye. Light passes through a series of transparent structures and humors before being absorbed at the back of the eye by a pigmented layer, which is part of the choroid.

---

8   I often begin this block with "a riddle" from Friedrich Schiller's *Turandot*. Can you solve it? (transl. M. Holdrege)

Can you name me the crystal,
To no other can it be compared,
It rays forth light without ever burning,
The cosmic all it draws within.
The sky itself is painted on it,
Its image on the wondrous ring.
And yet, what rays forth from it
Has more beauty than it taketh in.

Part I. 9th Grade

The students are often surprised at this. Who would have thought that the interior of the eye would be a pitch-dark, hollow space, filled with a transparent and colorless gelatinous mass (the vitreous humor) all the way from the lens to the retina? Only at its very periphery is the space lined with a fine nerve layer (the retina) and light-absorbing pigment—and, then, hidden behind them, a layer of blood vessels (the choroid), which borders on the outermost layer of the eye, the sclera. From the point of view of the light, the eye appears to be a large, open space that provides entry into the human body, and only at its very back is there a thin nerve layer (the retina), through which we are sensitive to the light's presence. (Even though, as we shall see, more anterior parts of the eye—the iris and the lens—already begin to react to and modify the light's entry based on where the observer's attention is being focused.)

After this visual overview, we can go back through the components of the eye and learn something of their unique qualities. An obvious place to start is the crystal-clear cornea, with its bulging anterior surface, which is an important part of the image-focusing structure of the eye. It consists of some 60 tissue layers that contain no blood vessels, and which must, therefore, receive their nutrients from without by diffusion. In order to maintain its transparency, the cornea requires the right degree of intraocular pressure from the fluids flowing slowly through the anterior chamber behind it.

Depending on the class, the teacher will vary the degree of detail covered when going through the eye's components in this first overview, but one fascinating fact that should not be left out is the makeup of the crystalline lens. It consists of some 20,000 thin concentric layers of transparent proteins (called crystallins) that are tightly packed together like layers in an onion. Since new lens fibers are added throughout life, the lens gradually becomes denser and less elastic. This has consequences—about which the students will soon learn. Another "fun phenomenon" not to be overlooked is the appearance of *mouches volantes* (flying mosquitos, vitreous floaters) in conjunction with the vitreous chamber.

After a brief characterization of the internal organs of the eye seen in our drawing, we have enough background to consider the process of accommodation. Here we find clear causal interactions that speak to the thinking of a 9th grader. I have found it helpful to begin with an exercise that demonstrates what we are dealing with. The students are asked to position one thumb about four inches in front of their eyes, and their second thumb about a foot in front of their face. They are asked to focus first on the thumb in the distance, at which point they suddenly "see" that they have three thumbs, not two! How can this be? And what's more, the two peripheral thumbs are fuzzy, while the one in the middle is clear. With the help of a drawing something like the one below, we explain how the double vision arises for the thumb that is not being focused on.

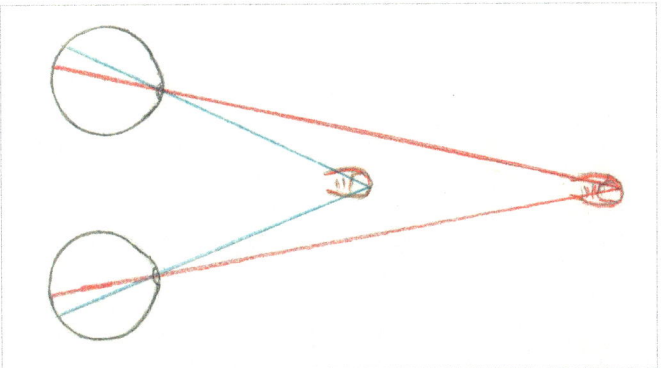

Fig. 1.3 Focusing on the distant thumb (the red lines lead to the fovea) creates one focused image of the distant-thumb, with two blurred versions of the near-thumb on both sides of it.

But why are the two versions of the near-thumb fuzzy and the distant one is not? To explain that we need another drawing that shows how when we are focusing on something, the image is refracted by the lens (and the cornea to a lesser degree) so that it gathers exactly in one small area of the retina, at the fovea centralis. There we have clear vision.

As the drawing shows, to look at something close up requires that the image be refracted more strongly than when looking farther away. On this basis, we can explain why we cannot clearly see both the distant and the close-up thumb(s) at the same time.

Once we have that figured out, the question becomes, how do we change the level of refraction to correspond to what we are focusing on? Since the lens is the main refractor, clear logic tells us that it must be involved and somehow changing to fit each new situation. After some discussion (speculation) about how this could be happening, the students learn—with the help of additional sketches—about the relationship between the elastic lens, the radial fibers (suspensory ligament), and the ciliary muscle.

Part I. 9th Grade

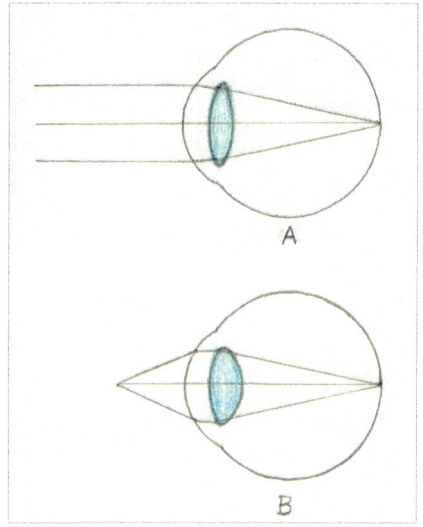

Fig. 1.4 A. Looking at a distant object: the lens is thin. **B.** Looking at an object close up: the lens thickens, thereby refracting the light more strongly.

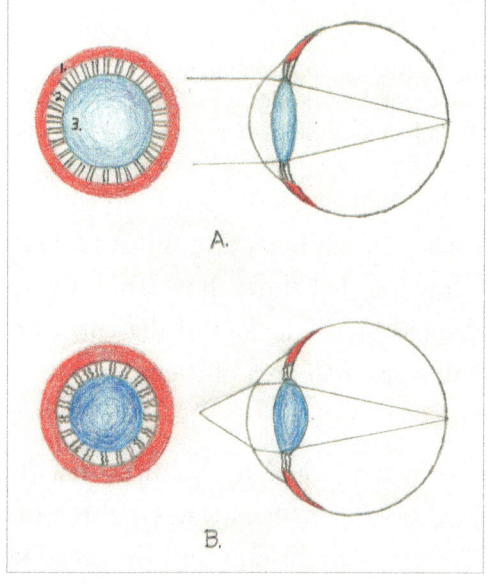

Fig. 1.5 1. Ciliary muscle. 2. Radial fibers. 3. Lens. A. Object distant: ciliary muscle relaxes and radial fibers are pulled taut, stretching the lens thin. **B.** Object close: ciliary muscle contracts, radial fibers relax, and the lens thickens through its own elasticity.

The students now come to the amazing realization that our eyes are able to constantly modify the thickness of the lens as we view objects at varying distances. Only at distances of roughly 200 ft and beyond is no such modification necessary.

So, using the student's growing thinking capacities, we can now move through a very clear chain of causality: Close-up objects need greater refraction to come into focus, which requires a thicker lens, which necessitates the contraction of the ciliary muscle, which causes a loosening of the radial fibers, which allow the lens to grow thicker through its natural elasticity, which enables it to refract the image so that it is focused at the fovea centralis. We reverse this sequence when looking at a distance and find: A thinner lens can be created by the pull of the radial fibers, which is caused by the relaxation of the ciliary muscle, and so on. And in this way, our eyes move back and forth between these two poles, determined by whether we direct our attention far or near.

Before we address the question of why images need to be focused on the retinal fovea and not somewhere else, we can ask why some people are unable to focus clearly at a distance, while others have problems seeing close up. To answer this question, we create sketches on the board that illustrate the nature of near- and farsightedness.

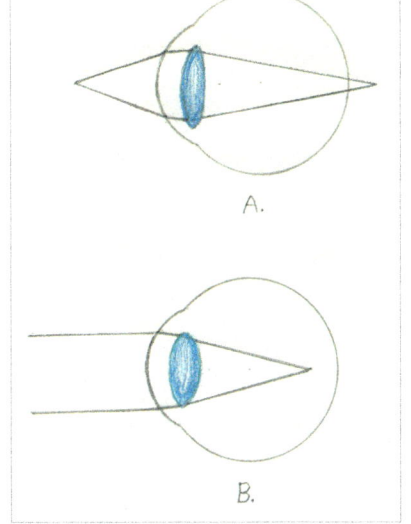

Fig. 1.6 A. Far-sighted: Image of close object not refracted strongly enough to be focused at the fovea. **B.** Near-sighted: Distant objects are refracted too strongly and thus the focus point falls short of the fovea.

If people have such issues, how do we solve them? With the help of supplemental lenses, of course! Further illustrations show how a convex lens serves the farsighted and a concave lens the nearsighted.

Next we can discuss the students' parents, most of whom now need reading glasses, but didn't when they were younger. What's up with that? The lens loses flexibility with age. If the lens can no longer thicken to the same extent when the radial fibers release their tension, it would lead—logically—to the inability to see things clearly close up. And the solution: a pair of convex lenses (also kn own

Part I. 9th Grade

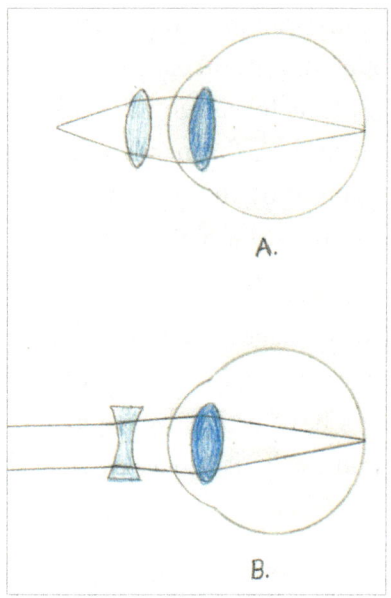

Fig. 1.7 A. Convex lens adds additional refraction, thus enabling the image to focus at the fovea. **B.** Concave lens disperses, thereby balancing out excessive refraction and allowing the image to be focused at the fovea.

as reading glasses)—easily available by the hundreds at your local supermarket, dollar- or drugstore.

It's time to gather some data on this. We measure the shortest distance that each student can see a 12-point font clearly and enter that data in a table. When they go home that evening they do the same measurement on their parent(s) and this data is recorded in the same table according to age groups (40–45, 46–50, etc.). The data speaks for itself: Despite considerable variation, it is clear that older people begin to lose some of their youthful flexibility (not only in their eyes, some students may wish to assert!).

So if focusing at a certain point in the retina (fovea centralis) is such a big deal for clarity of vision, it follows that logic-seeking 9th graders will want to know why?

A drawing of the retina (not shown here) can illustrate how in the fovea there are only color-sensitive cones, whereas rods lie further out toward the periphery of the retina.[9] So? Sharpness of vision is independent of color, is it not? Therefore, there must be more to it. Indeed, when we inspect the drawing

---

9   There are approximately 120 million rods per eye and 6 million cones—a 20:1 ratio.

The Eye and Seeing

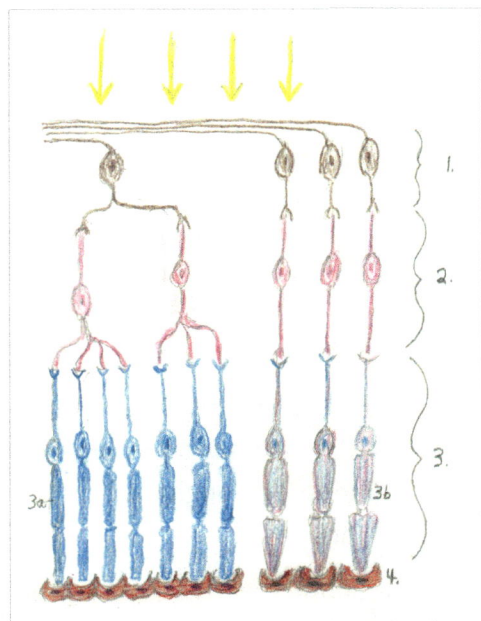

Fig. 1.8 Diagram of a small section of the retina. 1. Ganglion cells. 2. Bipolar cells. 3. Photoreceptors: 3a. Rods. 3b. Cones. 4. Pigmented epithelium.

above more carefully, we find that there is a difference in the way the rods and cones are connected to the other retinal neurons. Multiple rods are connected to each bipolar cell and multiple bipolar cells connect to each ganglion cell. In sum, as many as 100 rods feed into each ganglion cell. Their effects are thus summated and considered collectively. This is analogous to many voices speaking at once, which makes them easy to hear, but not very clear. This convergence allows the rods to provide high sensitivity to low levels of light, but at the expense of visual acuity. Each cone in the fovea, on the other hand, has its own individual bipolar cell that it connects to, which, in turn, connects with only one ganglion cell—a 1:1:1 ratio. Since the axons of the ganglion cells converge to form the optic nerve, it becomes evident that each cone has something comparable to a direct line to the White House (aka the visual cortex at the back of the brain).[10] All of this results in detailed, high resolution impressions of small areas of the visual field (Marieb & Hoehn 2012). So once again, we have found a way of explaining in a logical manner what at first appears to be a puzzling phenomenon.

---

10 Moreover, the indentation always found at the location of the fovea centralis in overview drawings of the retina is caused by the fact that the bipolar and ganglion cells there lean back and thus allow the light to shine past them and more directly onto the cones at their base. (It often helps to illustrate this with a drawing.)

Part I. 9th Grade

After learning about (and understanding) how the lens adjusts to correspond to changes in distance, we also consider how the colored iris—which (as we noted earlier) is located between the lightest and the darkest regions of our body—is also intimately involved in the play of light and darkness as it is received by the eye. The iris serves as a kind of guardian of the doorway to the eye's interior, by narrowing when there is excessive brightness, and widening when things grow dim. And so it plays a balancing role in our relationship to the ever-changing interplay of light and darkness. It also interests the students to hear that the iris widens when viewing our current "heart throb," or when we are amazed about something, whereas it narrows when we are afraid of or disgusted by something. We learn, too, how at a chemical level the eye adjusts to changes in light intensity through the breakdown and regeneration of the light-sensitive pigment rhodopsin.[11]

Yet another eye dynamic central to our visual experience is the coordination of the movement of our two eyes. We explore in thought—and with sketches if needed—what we think would happen if the two eyes were not coordinated in their movements, and conclude it would lead to double vision.[12]

But how do we move our eyes, actually? We reflect on how we move other parts of our body and conclude: Well, there must be muscles involved somehow, but where are they? Do they push or pull? Are they inside the eye or out? We know they are not inside (except for the ciliary muscle), so, logically, they must be outside. Where would a muscle have to be located that could cause the eye to rotate upward, for example? We take a basketball and experiment with this to see how movements to the right or left, up or down, etc., could be caused. This seems pretty straightforward, and when—finally—they get to see a drawing on the board of the six extrinsic eye muscles, it turns out that they had estimated pretty well.

---

11 For some classes it is also valuable to explore how complementary colors arise through a few experiments (focusing on a red circle for a few minutes and then looking at a white sheet of paper, for example) to literally "see" how the eye produces the complementary color to what it has been viewing. Here, again, we find a balancing gesture: producing the opposite quality and thereby overcoming one-sidedness.

12 The students love the story (fictional) about a boy whose father spoke to him about his concern that the boy might have double vision. The boy responded: "Come on, Dad, if I were seeing double, I would see four moons, not two! Duh."

The Eye and Seeing

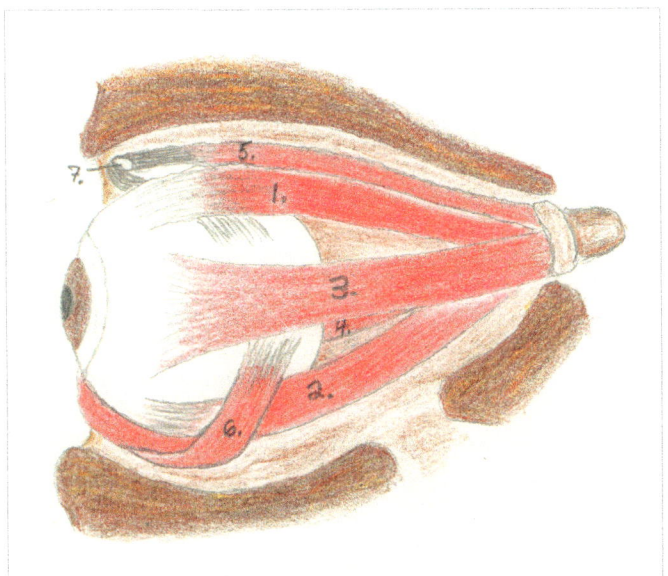

Fig. 1.9 The six extrinsic eye muscles. 1. Superior rectus.
2. Inferior rectus. 3. Lateral rectus. 4. Medial rectus.
5. Superior oblique. 6. Inferior oblique. 7. Trochlea.

We reflect on how exactly we are able to coordinate these movements and yet do so without knowing how. The teacher explains that there is even more significance to these eye movements than they could ever imagine, but that this will first be covered in a week or so in the context of another central human sense—one which most people have never heard of![13]

It is also valuable to give the students a sense for something that, although it transcends our attempt to reveal straightforward physical causality in the eye, is essential in every act of seeing (and in perception altogether). Yale philosopher N.R. Hanson famously put it into words this way: "There is more to seeing than meets the eye!" We discover this by looking at several "ambiguous figures" that can be viewed in more than one way.

It turns out that this is not a matter of changing images on the retina, but rather, as Hanson put it, of "the organizing activity of the mind" (Hanson 1958).[14] To explore this fascinating fact in more detail belongs in a 12th grade

---

13   Big hint: proprioception and saccadic eye movements
14   This is what Barfield calls "figuration" (see p.18).

Part I. 9th Grade

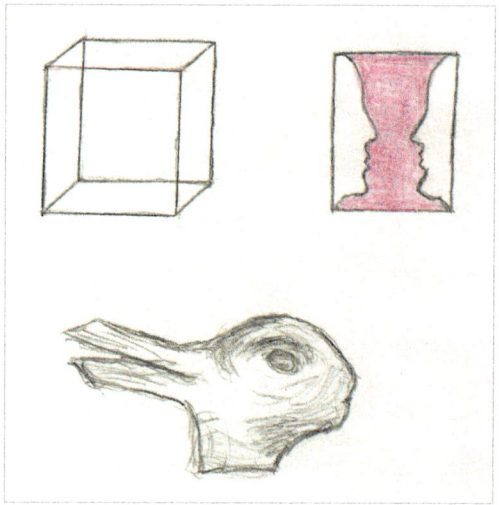

Fig. 1.10 "There is more to seeing than meets the eye."

philosophy class, but I think it valuable to give the students a small taste of this aspect of perception before they are ready to process it fully.

As we draw to a close in our study of the eye and vision, I have found it valuable to supplement the insights just gained about the "organizing activity of the mind" with a few very important anatomical/neurological facts. Contrary to the picture that most people have, the image created on the retina is not simply projected into the visual centers of the brain's occipital lobe. Instead, something very different takes place. There is a complete "deconstruction" of the image!

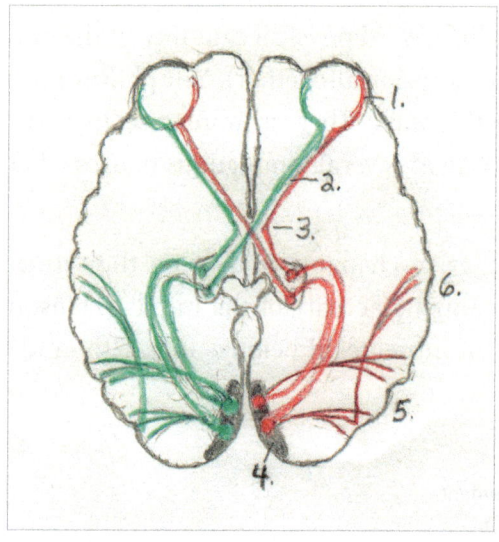

Fig. 1.11 Horizontal section of the optic tract leading to the primary visual cortex and to secondary and tertiary processing areas. 1. Retina. 2. Optic nerve. 3. Optic chiasm. 4. Primary visual cortex. 5. Secondary processing areas. 6. Tertiary processing areas.

One aspect is that the fibers from the left side of each retina go to the left side of the brain, and from the right side of each retina to the right side of the brain. Moreover, different elements of what was a coherent image (color, shape, space, movement) are transmitted as neurological processes to different (secondary and tertiary) visual centers of the brain. Humpty Dumpty is totally taken apart, and nothing of the original image remains at the neurological level (Rohen 2007, 1978).

It is also possible to supplement this insight with a comparison of the embryological development of the human eye with that of a squid. Both have highly developed eyes, but the human eye grows from the inside out—as an extension of the brain—with only the lens coming from the outer skin (epidermis), whereas in the squid—an animal whose behavior is generally directed outward and is tightly connected with its environment—the eye forms primarily from the outside in.

One effective way to round off the study of the eye is to "bring it down to earth" by describing a number of eye maladies.[15] The students are also interested to learn about Lasik surgery, something many of them have heard of, but which they now—based on what they have learned in this block—can understand in a new way.

---

15 Cataracts, glaucoma, astigmatism, strabism, etc.

Part I. 9th Grade

The Ear and Hearing

After studying the eye and seeing, we have a simple and obvious comparison point for what comes next: the ear and hearing. Comparing the two, the students immediately call attention to the significance of their differing locations. Although both are symmetrical pairs, the eyes are directed toward the world in front of us, where we are wide awake. The ears, on the other hand, are much more open to the entire surroundings. But sitting on opposite sides of the head, they are not able to focus in the same way that the eyes do. The next question becomes: Can we, in fact, focus our hearing? We try some listening experiments where the task is to focus in on certain sounds around us, and very quickly we have a definitive answer to that question. Interesting too, is to play the recording of a symphony and ask the students to shift their attention from time-to-time to different instruments and to follow them at the exclusion (more or less) of the others. The students also enjoy trying to imagine what it would be like if the eyes and ears were to switch positions.

Looking at the two organs externally we find a large contrast. Whereas the eye is a shiny sphere that we can look into, but which is otherwise closed off from the outer world, the ear presents a huge spiral-like veranda at the entrance to a tunnel that disappears into the side of the head.

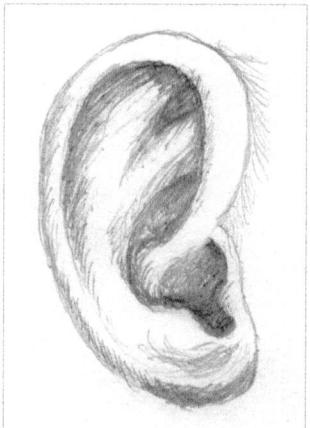

Fig. 1.12 The auricle of the human ear.

The outermost part of the ear, the auricle or pinna, varies from individual to individual, yet maintains the same basic elements and overall shape. When I ask the students if it has a functional significance, most conclude that it must assist in the gathering of sound. They cup their ears to see if enhancing the auricle in this way influences their hearing. It is fun to compare the human auricle with that of various animals—cows, dogs, horses, deer, etc.—which makes obvious that the human auricle is quite modest, barely manifesting what we find in many animals. But not only in shape and size. The auricles (pinnae) of many mammals are very dynamic and show where the animal's attention is focused at the moment. Some, like horses, can even point their auricles in different directions at the same time! (Multitasking?)

What would it be like to have an animal-like auricle, we ask. (Some students enjoy creating large cone-like auricles out of construction paper and to test how this influences their hearing.) The students are also amazed to learn how the masters of heightened hearing, the bats, are able, when they need to, to change the shape of their auricles from one extreme form to another in less than one-tenth of a second, which is two to three times faster than the blink of a human eye! (Gao et al. 2011)

Fig. 1.13 The auricles of a bat.

Even though we have six intrinsic muscles in each auricle, plus three extrinsic ones, only the most gifted of students (or so they may think!) are able to wiggle their ears at will. (If time allows, a wiggle completion can also be held, with the winner given the honor of cleaning the boards after class.) But the apparent modesty of the human auricle should not fool us. Researchers have found that its twists and folds are such that they enhance sounds that have a pitch typical of the human voice (around 3000 HZ). They enhance these sounds up to 100 times, while having no influence on pitches at other levels, thereby reducing background noise. Not surprisingly, they also support sounds coming

Part I. 9th Grade

from the front and sides of the head, while reducing those from the back (Science Buddies 2015). Such perspectives make us ready to explore just where that dark tunnel into the side of the head leads.

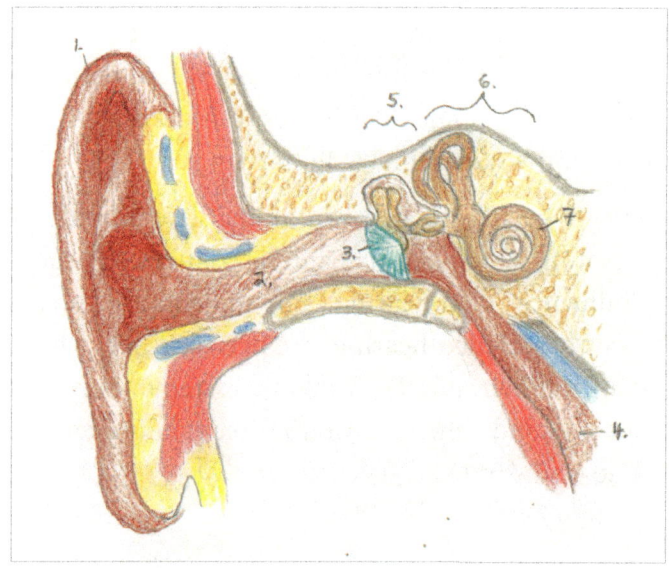

Fig. 1.14 Overview of the human ear (frontal section).
1. Auricle. 2, External auditory canal. 3. Ear drum (Tympanic membrane). 4. Eustachian tube. 5. Middle ear. 6. Inner ear.
7. Cochlea.

On the way in we find hair follicles and wax-producing glands along the walls of the auditory canal that help catch dust and other such contaminants that could hinder the vibrating of the ear drum. This does not mean, however, that sound is carried into the ear by moving air. Instead, it is the rapid rhythms (vibrations) of contracting and expanding air that spread outward from the source of the sound and enter into the external ear canal as sound waves.

That a drum-like membrane would resonate with these vibrations makes complete sense to the students, but what follows is more surprising. Connected to the eardrum is a tiny bone, the hammer (maleus), which connects to another bone, the so-called anvil (incus), via a saddle joint, followed by the stirrup bone (stapes), that is connected to the anvil by a ball joint. Where else do we find linear bones connected to each other by saddle and ball joints, we ask? In our limbs of course! With these three bones (the ossicles) we have something like a miniature leg kicking deep inside the human head (on both sides of it)! This

The Ear and Hearing

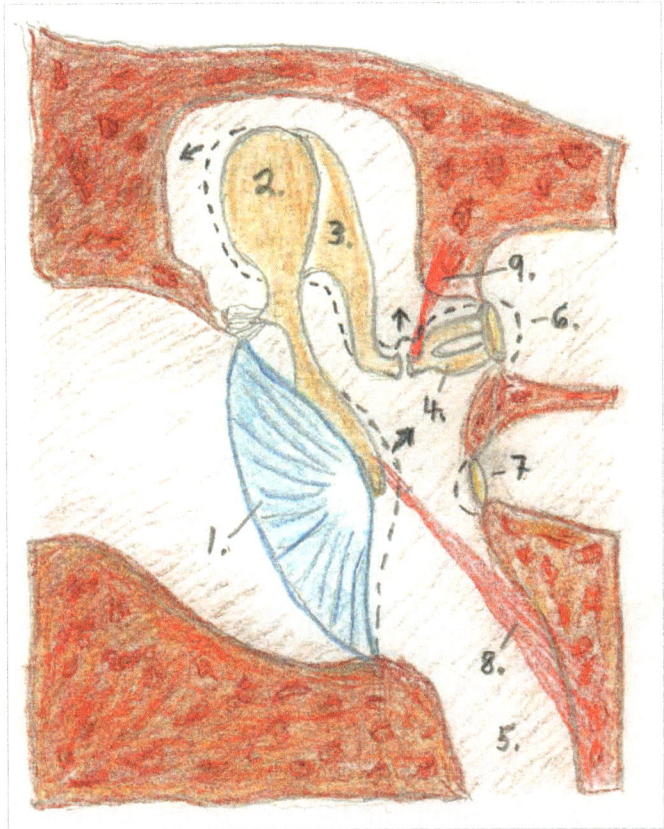

Fig. 1.15 The middle ear. 1. Ear drum. 2. Hammer (Malleus). 3. Amboss (Incus). 4. Stirrup (Stapes). 5. Eustachian tube. 6. Oval window. 7. Round window. 8. Tensor tympani. 9. Stapedius.

tiny leg is brought into motion by the ear drum, and at its far end (the foot-end) it "kicks" into the opposite wall of the middle ear, where the oval window is located. The oval window is a membrane that connects the middle ear to the inner ear (the cochlea). It passes on the sound waves from the kicking leg to the fluid (perilymph) in the cochlea. These waves spiral their way down to the center of the cochlea (more on that soon).

If we take a step back at this point—before considering the middle ear and the cochlea in greater detail—and consider this path of the sound waves from the source of the sound down into the heart of the cochlea, it appears that we have a straightforward causal chain of events by means of which the sound waves get passed on from one medium to the next: air → ear drum → ossicles →

oval window → perilymph fluid in the cochlea. This is just the kind of clear linear causality that speaks to the 9th grade mind.

However, if we reflect on this path of the sound from the point of view of maximum efficiency, which also makes sense to a 9th grader, then we can ask: Do we really need the middle ear with the ossicles. Couldn't the eardrum just pass on the sound waves directly to the cochlea without the big gap created by the middle ear and the complicated interaction of three bones? We've all learned since we were children about the amazing wisdom of the human body, but could it be that in this instance we are smarter and could build a more efficient system than the body itself if we just got the chance? This question we want to ponder over night. The next day it seems clear to most that we couldn't have outsmarted mother nature that easily. There must be more to the matter, something we haven't taken into consideration. If need be, the teacher helps the investigation along by suggesting the students compare the difference between creating waves through the movement of their hands in the air as opposed to under water.

With such an example in mind, it is clear that the transfer of the ear drum vibrations to the denser medium of the perilymph in the cochlea requires more force than it needed to move them through the air. Might this be the role of the ossicles? If so, how? At this point the teacher can offer one fact not readily available to the students: that the hammer and anvil are linked together in such a way that leverage is gained, thereby increasing the force that the stirrup can apply to the oval window. A second, decisive factor can be discovered by careful observation of the drawing of the middle ear. When one compares the size of the ear drum surface with that of the oval window, it is clear that the surface area onto which the vibrations are transferred through the ossicles has decreased significantly (60 mm² to 3.5 mm²). Does that matter? The teacher can ask for a courageous volunteer willing to test this. The student places their hand on the floor and lets one of the lightest students in the class place the broad heel of their shoe upon it, gradually increasing the weight applied until the student says stop. If done gently and gradually, their entire body weight can sometimes be applied. The teacher then offers the student a change of shoes, this time a finely-tapered shoe with stiletto heel.

Fig. 1.16 A stylish stiletto heel.

The experiment is then repeated with the new shoe. Usually this does not happen, however, because the courageous one knows their own limits and recognizes immediately that the same amount of weight applied through the new shoe would be infinitely more painful. The whole class sees this immediately. Why so, we ask? Reflecting on this together, it becomes clear that what had been a certain amount of pressure spread out over a large surface area is now concentrated onto a much smaller one, thereby increasing the pressure in that small spot enormously. Other examples of this soon come to mind. For the same reason, a sharp knife edge cuts more easily, darts stick to the wall, etc. In the case of the oval window, its much smaller surface area relative to the ear drum creates a great increase in the pressure that can be applied to the perilymph on the other side. In fact, the combined effect of the two factors—which we have named the "leverage effect" and the "stiletto effect"—yield approximately a 20-fold increase in pressure by which the sound vibrations can be transferred to the oval window and then on to the dense perilymph fluid found in the cochlea (Mörike et al. 2001).

So what happens to those sound waves in the cochlea? With the help of drawings, the students learn how the waves spiral two-and-a-half times into the auditory snail through the scala vestibuli. Reaching the center (the apex), the waves pass through an opening (helicotrema) and spiral outward again through a separate passageway, the scala tympani. At the end of this passage is the round window, a membrane that creates the boundary to the middle ear and which lies below the oval window.

Why a membrane here, we ask? Or, put differently, what would happen if at the end of the passage were just a solid "wall" similar to the exterior of the cochlea? It doesn't take long for someone to realize that the sound waves would

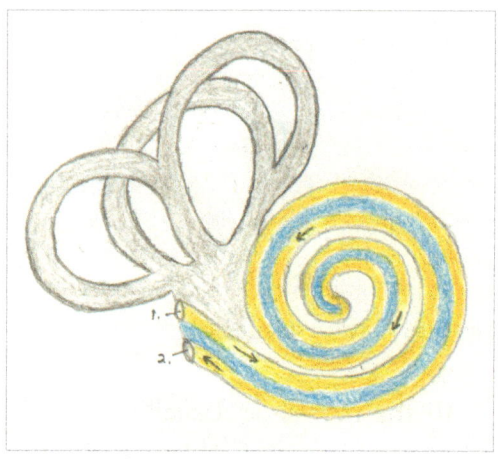

Fig. 1.17 Movement of perilymph waves into and out of the cochlea. 1. Oval window. 2. Round window.

bounce back, reversing their previous direction, thereby causing havoc with the new sound waves following them. The membrane of the oval window, by contrast, absorbs the waves and prevents such an echo-effect from happening.

We now look at the flow of the waves in more detail with the help of a vertical section through the cochlea, which shows that it is actually made up of three scala (passageways) that are separated by two membranes, the vestibular and the basilar.

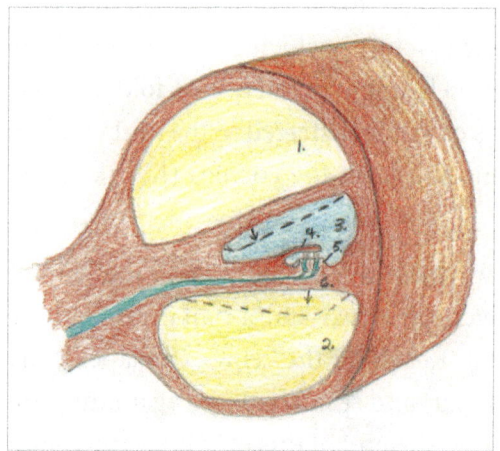

Fig. 1.18 Vertical section through one turn of the cochlea. 1. Vestibular canal (scala vestibule). 2. Tympanic canal (scala tympani). 3. Cochlear canal (scala media). 4. Tectorial membrane. 5. Organ of Corti with hair cells. 6. Basilar membrane.

The students learn how the waves passing down the scala vestibuli create pressure waves in the endolymph of the cochlear duct (scala media), which are transmitted to the organ of Corti, which is located on the basilar membrane.

The Ear and Hearing

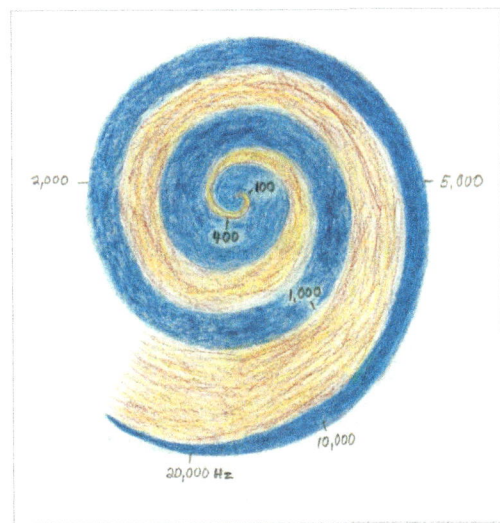

Fig. 1.19 Diagram of the cochlea from above showing that the fibers of the basilar membrane are "tuned" to particular sound frequencies, from 20,000 Hz near the round window to 100 Hz near the center of the spiral (helicotrema).

The fibers in each region of the basilar membrane vary in length—the shorter ones are nearer to the oval window, the longer ones toward the center. In correspondence with their length, the fibers vibrate with maximum amplitude to sound waves of different frequencies. Sound waves of a higher frequency (pitch) cause maximum basilar vibration closer to the stapes, lower frequency pitches cause peak displacement closer to the apex. The sensory hair cells located in the organ of Corti are embedded in the tectoral membrane that overhangs them (see drawing). They bend when the basilar membrane vibrations peak in their respective regions. This bending of the hair cells triggers specific stimulus patterns in the auditory nerve, which go to the auditory cerebral cortex of the temporal lobe. Neurons in different regions of the basilar membrane stimulate neurons in different regions of that cortex, where each region corresponds to a different pitch.

As the spiral-shaped diagram of the cochlea seen above shows, the normal range of hearing for a human being lies between the high tones of ~20,000 Hz, which are perceived at the base (beginning) of the cochlea, and the low tones of approximately 100–200 Hz which are registered at the apex. One obvious comparison to this organizational principle are the strings of a piano. There we have the high notes (and the short strings) at one end of a piano keyboard, which then gradually transition all the way down to the low notes at the other end. Unfortunately, as we age we begin to lose our ability to hear the higher tones. (That is why we plead with the students not to apply this newly-acquired

Part I. 9th Grade

knowledge by whispering to each other at high frequencies so that their older teachers will not be able to hear them!).

The students also like to reflect on how the range of hearing in animals compares with our own. We let them speculate for a bit about which animals are particularly attuned to high or low frequency tones before revealing what researchers have found. Not surprisingly, there are considerable differences between animal species:

Approximate Hearing Range (Hz)

human	64-23,000
dog	67-45,000
cat	45-64,000
cow	23-35,000
horse	55-33,500
sheep	100-30,000
rabbit	360-42,000
rat	200-76,000
mouse	1,000-91,000
gerbil	100-60,000
guinea pig	54-50,000
bat	2,000-110,000
elephant	16-12,000
porpoise	75-150,000
owl	200-12,000
chicken	125-2,000
beluga whale	1,000-123,000

https://www.lsu.edu/deafness/HearingRange.html

Once the above data has been revealed, a lively discussion among the animal lovers in the class often follows. (Some conclude—perhaps too quickly—that the best way to woo their cows and elephants is with low, mellow tones, while the mice that share their bedrooms should be greeted in high, squeaky pitches.)

At this point it makes sense to step back and consider what we have actually come to so far. On the one hand we have been able to follow the path of sound waves in a quite straightforward manner from the sound source into the spiraling cochlea. But to find out that at the end of this passage there is an unbelievably sophisticated and finely-tuned piano-like organ that is only 1/3 of an inch across and 1/5 of an inch high is quite amazing! The cochlea is about the size of a split pea and is embedded deeply in the human skull, in the hardest bone of the human body: the petrous bone.[16] It is incredible. We are able to listen to an unending variety of the sounds around us, some of them far off in the distance, and we do it by means of this tiny, tiny (miniature) organ, that is deeply implanted in the densest of bone in the human body located just below our eye sockets. Wow! (Marieb & Hoehn 2012, Rohen 2007)

During our explanation of the ear and hearing, several additional aspects that lend themselves to clear causal explanation can be looked at along the way. They can be easily covered when the middle ear and the ossicles are discussed. When drawing the ossicles, we can take note of two muscles in the middle ear that would seem to play a role there: the tensor tympanii, which connects to the base of the hammer, and the stapedius muscle, which connects to the stapes (stirrup). They present somewhat of a riddle, however, when we consider that the three ossicles are not brought into movement by muscles—as is elsewhere always the case!—but instead by the vibrating ear drum.

So what are these muscles good for if they aren't causing the bones to move? After some discussion—which may or may not come close to a satisfying solution—the teacher can describe what happens when you have been at a loud concert (maybe even a "rock concert") and discover when leaving that you seem to be going deaf! But, thank goodness, after a few minutes you are able to hear well again. What's up? Very loud sounds can be damaging to the ossicles and cochlea. The ear, however, reacts immediately in their presence with the contraction of the two muscles just mentioned, thereby hindering an overly high intensity (amplitude) of vibration in the ossicles and protecting the ear—as well as providing us with a more pleasing and moderate sound experience. (One reason why explosions can be so damaging to our hearing is that the middle ear muscles are not able to react in time to minimize the intense vibrations.) After

---

16   The petrous bone, in contrast to other bones, does not develop further after birth. It contains no bone marrow, is highly calcified and "unusually dead"—thus its name "petrous," which means "stony" (Rohen 2007).

the band stops playing, it takes some time before the two muscles can relax again, and until then, we don't hear very well (low amplitude). Once again, clear causality. And also a little bonus: We have discovered muscles that don't cause movements but hinder them!

The middle ear is also interesting to the students because understanding how it reacts to changes in air pressure can explain with great causal clarity something they have all experienced many times but have never understood. They now get why their ears go "pop" (after a short period during which they can't hear very well) when they go up (or down) a ski lift or elevator, drive up (or down) a mountain pass, or sit in an airplane as it climbs into (or descends from) the sky. The teacher can also shed light on how and why yawning, burping, and even chewing gum, help facilitate the pressure-equalizing process that is behind these phenomena.

Classroom experiments where the students try to discover how well they can sense the direction from which a sound is coming are also fun and informative. As it turns out, humans are normally able to tell the differences in the direction that sounds are coming from if they arrive at an angle of 8.5 degrees or more. (Dogs can sense differences at an angle of 2.5 degrees, cats at 1.5 degrees). How do we do this? We are highly sensitive to the time-difference in the arrival of a sound at each of our two ears. We are able to sense arrival differences as small as 0.00003 (3/100,000) of a second! (Mörike et al. 2001)

In conclusion, it is often worthwhile to have the students reflect on and discuss how different it is to be deaf, as opposed to blind. Very interesting observations can arise in such discussions. Some students also appreciate the opportunity to ponder this question more deeply and to write a short essay on where their reflections on this topic have taken them.

The Sense of Balance (Equilibrium)

After exploring the inner ear, the students are curious to find out about the three tube-like canals that lie above it and appear to be part of the cochlea. But before that organ is addressed, it is helpful to gather some experiences that can lead us into this theme. A good starting point is to have the students stand with their eyes closed and sense the subtle movements they make in their feet (rocking back and forth, etc.) to keep the column of their body from tipping over. When they stand on one leg (eyes closed) they can experience how much of the body gets involved in staying balanced. We gather further observations about which parts of the body were particularly active in maintaining balance. It is amazing how many ways that we engage our body in order to stay upright! It is also worthwhile to notice what we experience when we actually lose our balance and we have to put our other foot on the ground to keep from falling.

After such experiences, I often ask the students to stand and spin in a clockwise direction. They do this until they are about to fall and then stop. They notice (and remember from their younger days) that their surroundings keep moving even though they have stopped. I ask in which direction the movement kept going and all agree. Then they reverse directions and find that when they stop everything keeps moving, but in the opposite direction.[17] Then we discuss what it is like for a person who gets seriously dizzy through such spinning—something most of them have experienced once-upon-a-time at an amusement park. In extreme cases the person falls down, their heart begins to beat rapidly, they break out in a cold sweat, and they may even begin to vomit and get visited by diarrhea (the mention of which usually elicits an "ugh!" from some corner of the room).[18] This extreme reaction gives pause for reflection. Do other senses

---

17 An interesting question at this point is: How do professional ice skaters and dancers prevent this from happening? Invariably, several of the students will know the technique they use and will be able to explain (even demonstrate) it to the rest of the class.

18 If the mood is right, I sometimes tell them about the time I crossed the English Channel on a ferry during a major storm. It caused the ship to rock back and forth (to tip down in front 60° below the horizontal and then up in front 60° above horizontal, back and forth), with the consequence that the floors of the common area were soon covered with and sloshing back and forth the half-digested remains of many passengers' midday meal—in perfect synchronization with the rocking ship. (I only survived by lying down on a couch facing inward, with my eyes closed and holding my nose—both the sight of it and the smell were overwhelming.)

have such a massive effect when taken too far? Taste and smell can cause strong reactions, but are not normally so overwhelming—and they are triggered when something foreign to us actually enters the body. The class soon realizes that to have such a enormous effect, our sense of balance must be deeply interwoven with our entire organization.

Assuming our normal state of being in balance, we can also ask what it is that they perceive with this sense. Gradually it dawns on the students: This sense actually enables us to perceive the center of the earth! It puts us in tune with the midpoint of the planet—which is almost 4000 miles away! In other words, it establishes our relationship to the gravitational field we are embedded in. It engages us in a constant process of aligning and harmonizing of our bodily organization with that field of earthly forces. (But if we get out of sync there is "…ell to pay," as our foregoing discussion made clear.)

Another interesting phenomenon to discuss is how people react to "upside-down glasses"—glasses that invert the field of vision 180 degrees? With such glasses on, the class, the teacher, the desks, etc., all appear to be hanging from the ceiling, and we pour our morning cup of tea, it appears to be flowing upward! To get a sense for this, we all stand up and bend over so that we are now looking backward at a classroom that is downside-up (upside-down). Could we live our lives with a visual field like this, one that contradicts where our sense of "down" is. The students are very surprised to hear that people in experiments that required they wear upside-down glasses non-stop (except when sleeping) only saw the world that way for a period of time (Ahrahams 2012). Amazingly, after a week or so, these people—with their glasses on—could again see everything like they normally would (downside down). But, the bad news was, if they took off the glasses, everything would flip (turning downside-up) for them and they would need another week before they readjusted. What is happening here, we ask? After some reflection it becomes evident that when sight and balance are put in opposition to each other (contradict each other), one of the two must "give in" so that the two are in sync again. Which one proves to be dominant? Our sense of down! This tells us in yet another way what a central role balance plays, not only within our own organism, but also in our relationship to the outer world.

It is also worth considering that when we are running, jumping, doing gymnastics—activities where we are immersed in a sea of movement—we still maintain our sense of being centered, of having a center of gravity around which everything else revolves. It is amazing to run through a bumpy landscape, where visually everything is rising and falling, moving this way and that, and yet you feel yourself to be at the stable—even though moving—center of it all and in charge of what is going on—unless you lose your balance, start stumbling and tumbling![19]

The students also enjoy reflecting on what it is like for young children when they first stand upright, learn to walk, etc. (They love it if the teacher can give a lively—maybe even a bit goofy—imitation of this.) It is a uniquely human capacity. It is not instinctive, but is based on imitation and must be learned and practiced. (Later in the block we will see how becoming upright brings about significant changes in the structure and shape of various organs in the body.)

Returning to our spinning exercise, where the whole world kept rotating around us after we had stopped, the teacher can ask if the students know of other instances where things keep moving on their own without some obvious force propelling them ahead? Obvious examples that come up would be: A car keeps rolling when you take your foot off the gas; a baseball keeps flying after it has been hit; a hockey puck keeps pucking after being puckered, etc. At another level, satellites seem to keep going on forever; supertankers already turn off their engines 15 miles before they reach the dock; when moving at 55 miles an hour, the average freight train needs a mile or more to stop after the engineer applies the emergency brake full force. Why do we need seatbelts? What causes whiplash? What happens with your upper body when you jump from a train and your feet hit the ground first? With such examples in mind we can elucidate the concept of inertia.

With this conception firmly in place, we turn to the structure and functioning of the three semicircular canals aided by several drawings. We describe how the three canals are oriented at right angles to each other in the three planes of space, similar to the way that the three sides of a cube each have a different spatial orientation.

At the base of each canal is a bulge, the ampulla, which contains sensory hair cells embedded in a gelatinous membrane, the cupula. The cupula projects

---
19   If time allows, the teacher can go outside with the class and try this out.

Part I. 9th Grade

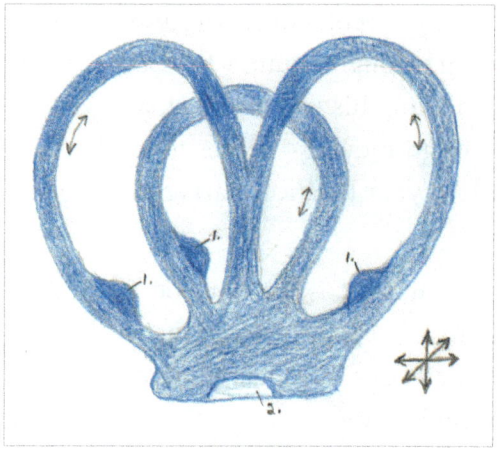

Fig. 1.20 The three semicircular canals, each with an ampulla at its base. 1. Ampulla. 2. Utricle. (Saccule not shown)

Fig. 1.21 Section through an Ampulla. 1. Endolymph. 2. Cupula. 3. Hair cells.

into the ampulla so that when endolymph flows through the canal it bends the cupula in the direction it is going, which creates a nerve impulse that is sent to the brain.

With all this as background, the students are now ready to see if they can figure out how what has just been discussed manifests in a hands-on example. Let us say that we rotate our heads to the side very quickly. What is happening in the three semicircular canals? The arch of each canal is located in one plane of space: front-back (anterior-posterior), side-side (horizontal), or up-down (vertical). Depending on the direction of the head movement, the endolymph in each canal will be affected differently. So if we turn our heads sideways, it will most strongly influence the flow in the canal that is oriented in the side-to-side (horizontal) plane. So what happens to the fluid in that canal?

As the students try to picture this concretely, they realize that once again, the concept of inertia is applicable. The static inertia of the endolymph will cause it to lag behind the movement of the head at first. This will bend the cupula in a backward direction relative to the head movement. But once the head stops, the dynamic inertia of the now moving endolymph will keep it flowing forward in the canal for a short time after the head has stopped, which will bend the cupula forward this time. The bending of the cupula stimulates the hair cells in the cupula, which activates sensory nerve cells that transmit impulses to the brain, which allow us to become conscious of the onset of the head movement (cupula bends back ← briefly) and when it stops (cupula bends forward → briefly).

So what happens if we spin around like we did recently and everything still seems to keep moving after we stop? What happens to the endolymph in this situation? Wow, the students realize, after a short lag-time it gathers so much momentum (dynamic inertia) that when the head stops turning the lymph keeps flowing and continues for a time to bend the cupula forward.

Here again, an enigmatic experience has been explained with straightforward causality. Of course, there are always deeper mysteries if we dig further, but we don't, because the focus of such investigations in the 9th grade is on a level of reality that can be grasped transparently through straightforward linear causality. In this way the 9th graders develop confidence in their thinking and its capacity to understand the world they have begun to awaken to!

But does this explain all the phenomena we considered earlier? What about a stable head tilt to the side, or when we experience an elevator starting or stopping without any visual cues? Here, too, we are dealing with gravity and inertia, but the ability to perceive such positions or movements (know as static equilibrium) requires a different organ structure and placement than the ampullae in the three curved canals. In the case of the elevator, it requires an organ structure that is sensitive to linear (straight line) changes in speed or direction (rather than rotational); in the case of the head-tilt, a structure is necessary that is sensitive to static positioning.

Part I. 9th Grade

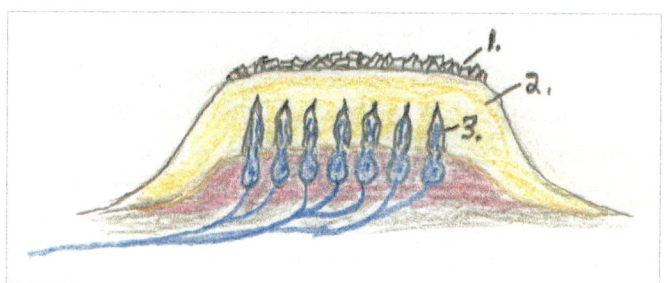

Fig. 1.22 Section through a Utricule. 1. Otoliths. 2. Gelatinous membrane. 3. Hair cells.

With the help of drawings like the ones above, we investigate another aspect of the inner ear, the maculae. One macula is located in the saccule wall of each ear and one in the utricle wall. The students learn (in more detail than I give here) that the maculae hair cells are embedded in a gelatinous mass that is studded with tiny calcium carbonate stones, the otoliths. In the utricle the macula is horizontal, in the saccule nearly vertical. We discuss the significance of having little stones embedded in the top of the gelatinous mass, how they could affect inertia and hair movement/positioning when an elevator takes off, or when the head is tilted. Once again, a clear causal explanation will emerge!

In this context, I often read to the class a humorous (but tragic) story from *The Man Who Mistook His Wife for a Hat* (Ch. 7), by Oliver Sacks, about a man who had lost his sense of balance and created a level that dangled in front of his glasses to perform that function for him.

The Sense of Movement
(Kinesthetic Sense, Proprioception)

After noticing how centered we are through the sense of balance, it is time to explore the opposite pole. We are beings whose bodies are in constant movement and in the most varied—yet normally very coordinated—ways. How do we manage this?

First off, we do a few simple exercises standing up. We touch the tips of our index fingers over our heads (without looking up) and then do the same behind our backs. We close our eyes and put our left pinky in our right "external auditory canal." We can sense very exactly whether a finger is bent or stretched, whether our leg is bent under the chair or stretched out in front of it—all without looking. Studies have shown that we can sense a change in the angle of our elbow equivalent to 0.038 degrees (Koenig 2006)![20] And this is almost insignificant when we consider the complex patterns of movement that we have to be in touch with when we are active athletically or when playing a musical instrument—not to mention speaking or singing.

We ask ourselves: How are we able to do all of this? Discussing this with the students, it becomes clear quite rapidly that the sense of touch is not sufficient, nor do we taste, smell, hear, see or balance where our limbs are or what they are doing. So, thinking logically, we conclude: Well, if we are sensing something, and the classic five senses plus our sense of balance are not able to do that sensing, then we must have more senses than we are usually aware of—more than the classic five and balance! We then try to imagine what such a sense organ might be like. Certainly not like eyes or ears, but couldn't we have something like tiny utricles or ampullae in our limbs? This doesn't make sense either, because we can twirl our arms for minutes at a time without experiencing anything like a continued spinning of the arm once the movement has stopped. Inertia doesn't seem to be at work here. And besides, the utricle and ampullae are embedded in bone and isolated deep in the normally resting head, where the body moves the least. Such organs would "go crazy" if located in areas that are constantly

---

20   Karl Koenig's book *A Living Physiology* (2006) is a wonderful resource for understanding the sense of movement being discussed here.

making the most varied motions with great rapidity and sudden changes. No way, we conclude. These organs must be very different from what we have seen up to now.

So we stop comparing these unknown structures with ones we know and focus on where they would need to be located. Simple logic tells us, if we are sensitive to movement all over our body, then we must have many such organs that are located wherever we are able to sense our movements. Simple logic also tells us they must be small and well hidden, otherwise we would already know about them just as we know about our eyes and ears.[21]

At this point (or in some instances earlier in the session after the movement exercises) the students are prepped to hear about an individual who lost this form of sense perception. They are fascinated and also touched when they hear the story of Christina, a "strapping young woman of twenty-seven," who awoke one day and couldn't feel her body, found that her hands "wandered" when she wasn't looking at them, and so on. Instead of just telling Christina's story, I also read extensive passages about it as told by Oliver Sachs in chapter three of his book, *The Man Who Mistook His Wife for a Hat* (1992). Not only do the students experience empathy for such an individual, but they also learn how she, with resolve and strength of will, learns to live with this condition and rebuild her life. In the process they begin to realize what an amazing capacity our sense of movement (proprioception) is and how we normally just take it for granted.

A great exercise that can follow a discussion of Christina's condition is to give the students a short (rather rapidly-spoken) dictation, which they write down using their non-dominant hand. They are amazed how difficult this is—and enjoy their incompetence greatly! We reflect on this, and realize: Although they know exactly what to do (how to form the letters)—that is not the problem—they don't have enough subtle awareness and control of their hand/finger movements to write with anywhere near the speed or elegance that they take for granted with the other hand. They experience how their movements do not follow their intentions with anywhere near the accuracy that they expect.

---

[21] It often makes sense to break up the class into small groups to let them discuss the questions just referred to. We can then let each group report its findings/conclusions regarding location and size of this mystery organ.

We talk about how much they practiced in grade school—for years!—in order to gain the kind of control over their movements that they simply take for granted as 9th graders. We talk about other forms of schooled movement, in the arts, in sports, etc. They realize: We have an amazing potential for the most varied and precise movements, but they must be acquired and practiced, often for years, and in some instances, life-long![22] After all, how long does someone like Yo Yo Ma practice every day?[23]

After various considerations of the sort just described, we turn next to the organ in question. With the help of drawings we learn about muscle spindles (proprioceptors) and similar receptors located in our skeletal muscles, in the tendons at the ends of the muscles, and in our joints.

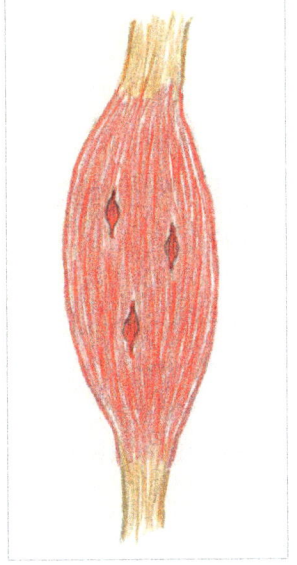

Fig. 1.23 Example of how muscle spindles can be embedded in a muscle.

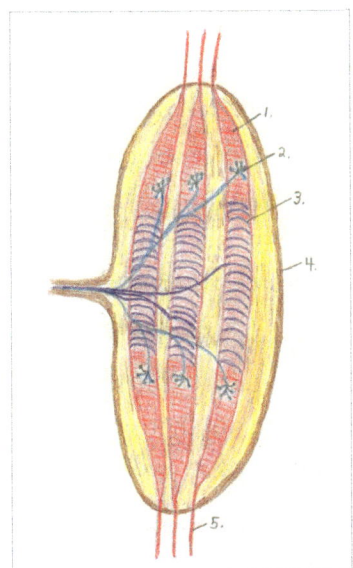

Fig. 1.24 Muscle spindle.
1. Intrafusal muscle fibers.
2. Secondary sensory endings.
3. Primary sensory endings.
4. Connective tissue capsule.
5. Tendonous fiber endings.

---

22 We will return to this topic in the 10th grade when we study the brain.
23 He estimates that he practices 2000 hours per year. (Do the math: an average of almost 6 hours per day!)

The muscle spindles located within the skeletal muscles are usually 2–3mm in length, and contain three to six tiny, embryonic-appearing striated muscle fibers enclosed in a protective sheath (intrafusal fibers). Since the spindles are arranged in parallel with the muscle fibers and interwoven with various nerve axons, stretching a muscle causes the spindles embedded within it to stretch, which stimulates the nerve endings in the spindles. (One could go into much more detail regarding the anatomy and functioning of the muscle spindles, but for 9th graders, an uncomplicated overview is sufficient and less confusing.) The receptors in the tendons continually monitor the tension in the tendons, whereas the proprioceptors in the connective tissue capsules of synovial joints are stimulated by the movement of those joints.

An obvious question that follows is: Are muscle spindles more numerous in some muscles than others and, if so, why? After discussing where it might make sense to have more spindles to enable more subtle movement perception, the students are not surprised to learn that they are very densely distributed in finger muscles and in the muscles associated with speech (more on that tomorrow!). Somewhat surprising, however, is the high density of muscle spindles found in the six extrinsic eye muscles. This is a highly important phenomenon and one that we will explore further.[24]

From the above we now realize that our muscles are not only a means for movement, but contain within them sense organs that perceive those movements. These are sense organs that we are normally unaware of, even though we are unceasingly immersed in and in need of the perceptions they provide!

But before we leave this topic, we must explore further why the eye muscles are so densely "spindled." We do an exercise with a red (or blue, or green) circle. At first we focus on the form as such (is it perfectly round, for example?), and after a bit we switch our attention to the color alone. Some students will notice that when we focus just on the form, we are more distanced, whereas when we concentrate on the color by itself, we seem to enter into the image much more—a kind of merging-with, rather than a more detached, wakeful assessing.[25]

---

24 The only muscles that don't have muscle spindles are those in the middle ear, about which we learned about a few days prior. Interestingly, those muscles, as we discussed then, are the only ones in the body that don't cause movement but actually dampen it.
25 See Rudolf Steiner's *Study of Man*, lecture VIII.

Next the teacher can unveil a quite abstract, two-dimensional form on the board and give the students a very quick look before covering it up again. They are asked to redraw the form the best they can from memory. This is not easy, so they get another chance to look at the form, which is uncovered for 15 seconds this time! Afterward, we discuss what they actually did to impress the form on their memory. Some notice how they retraced the angles and curves carefully with their eyes, and sometimes even used a sketching hand movement to get a better sense of the form. We talk about how one of the best ways to remember a form is to actually draw it. What sense joins our vision in such instances? The sense of movement, of course! We can take this dawning realization one step further by learning what happens when we look at a human face.

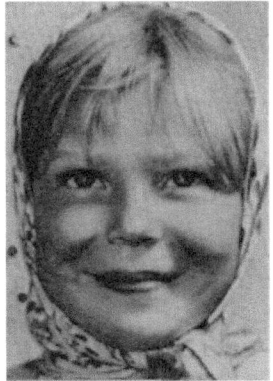 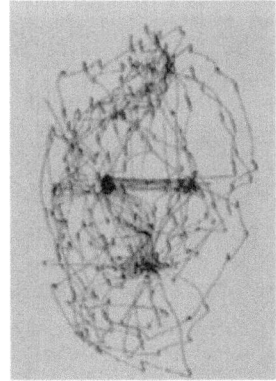

Fig. 1.25 The sketching movement of the eyes (saccadic eye movement) when viewing a human face (recorded by Alfred Yarbus).

When we observe an object such as the face above, we actually "sketch" its form(s) through small, rapid (saccadic) eye movements. The lines to the right of the face show the path of the eye movements (the dots represent short stopping points). Our eyes change their gaze-direction many times per second, and it is only through these "sketching" movements that we are able to experience the form-aspect in our field of vision.[26] In experimental settings where the eyes of subjects are not allowed to move in this way, those individuals lose their ability

---

26 One demo that the teacher might want to do, is to hunch their shoulders up around their neck for a period of time while teaching in front of the class and then ask if any of the students notice how their own shoulders are also beginning to feel tight and uncomfortable.

to see the forms around them, the images they had been viewing disappear. (Yarbus 1967, Kahnemann 1973, Julius 1984, Kranich 2003, Tatler et al. 2010).

It is, then, our sense of movement that allows us to experience the forms of things. We perceive the movements of our six extrinsic eye muscles when we are looking at things and through that experience the forms. That we have so little experience of the eye movements as such, has to do with something we learned about when discussing inertia. Because the eye is cushioned within the eye-socket (orbit) by pads of fat, it is basically "floating" (almost weightlessly) therein and can thus move essentially resistance- and inertia-free within that space. Due to this, it takes so little effort to move our eyes that we hardly notice this activity, but attend, rather, to the overall patterns that arise when we "sketch" the forms around us. If it took a lot of effort to move our eyes, we would put much of our attention there rather than on what we are looking at.

This can be made more tangible by asking a student to stand in front of the class and to observe an object or drawing in the back of the room. The others can see clearly the active eye movements that the student is making, even though the student herself is hardly aware of this, or may not notice it at all.

Summing up, the students are usually quite taken by the realization that, without our sense of movement (sense of form), we would not be able to see the objects in our environment clearly, nor would we have the movement capacities we take for granted. We would live, unmoving, surrounded by a sea of color! (Some students wish, nonetheless, that they had the option to experience the world in that way—sans our sense of movement—from time to time.)

The Larynx and Speech

After becoming familiar with the sense of movement and recognizing what a central role it plays in the relationship we have to our body and in our capacity to experience forms and movements around us, we can discover how this sense is engaged at yet another level: in our ability to express our inner life through speech. For, as we shall see, the central organs of speech depend intimately on our sense of movement.

We begin with a few voice exercises (consciously speaking several vowels, consonants, short phrases) that provide some basic observations about the process of voice production and its shaping. The flow of air from the lungs, to the larynx and further up in to the mouth-head region is quickly evident. We focus on the larynx and see if we can determine how its movements play a role here. We can also compare the size of the larynxes in the class and see if there is a correlation between size and tone of voice. After various explorations of this sort, several drawings help us further.

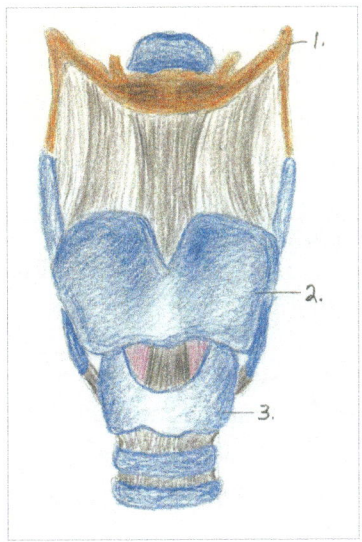

Fig. 1.26 Frontal view of the human larynx. 1. Hyoid bone (tongue bone). 2. Shield cartilage. 3. Cricoid cartilage.

As the image above makes evident, the larynx is a quite fascinating structure. The teacher will need several illustrations from various perspectives in order to make the actual shapes of the various components clear. It requires focus and considerable "perceptual intentionality" for the students to organize the parts in their minds ("figuration") to the degree that they can reach clear causal conclusions about how it works!

An obvious place to begin this investigation is the tipping back and forth of the shield (thyroid) cartilage, which we noticed through our own observations. A pivot point at the juncture of the shield and cricoid cartilages must be determined and then the muscles that cause the movement. Clearly, an attachment behind the pivot point (the ring and shield cartilage joint) would tip it back, and one (or more) in front would be needed to pull it forward again. Do they exist? Why of course, they must, our clear causal thinking tells us. And indeed, what anatomists have found is one muscle connected to each anterior side of the shield cartilage, which tip it forward, and a posterior muscle on each side that pull it back and downward.

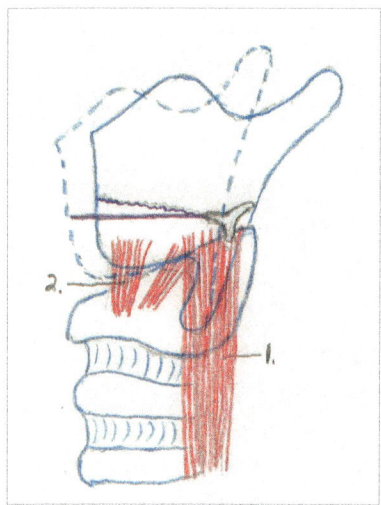

Fig. 1.27 Muscles that cause the shield cartilage to tip back and forth.
1. Sternothyroid muscle 2. Cricothyroid muscle.

Discussion about how this movement of the shield cartilage might influence the voice leads to the conclusion that a tipping back of the shield cartilage would cause the vocal cords to stretch and tighten, which would modify the pitch of the voice. This corresponds to our earlier voice experiments, when we could feel with our hands how the larynx rose and fell as we alternated between higher and lower tones.

A top-down (superior) view into the larynx reveals several structural features, one key one being that the upward flowing air must pass between the vocal cords. We also notice that the vocal chords can be drawn closer together (glottis narrow) or moved farther apart (glottis wide).

The Larynx and Speech

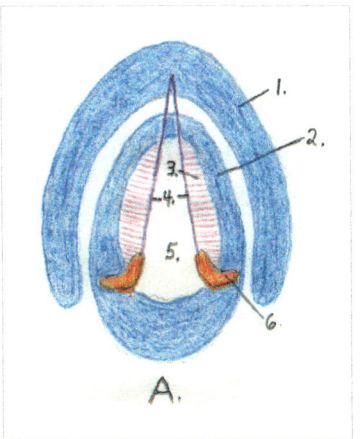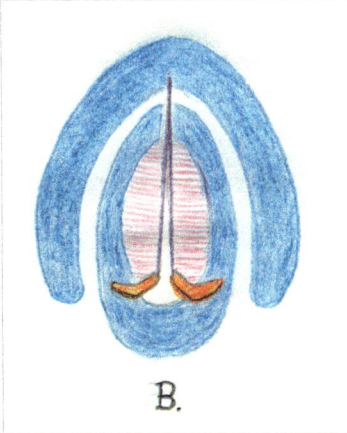

Fig. 1.28 Superior view of the larynx. **A.** Glottis wide. **B.** Glottis narrow. 1. Shield cartilage. 2. Cricoid cartilage. 3. Conus elasticus. 4. Vocal cords (vocal folds). 5. Glottis. 6. Arytenoid cartilage.

We discuss what effect "closer together" would have on the air flow—and other questions of this nature. Once the significance of such modifications becomes clear, we ask: How are the opening and closing movements of the vocal cords brought about? We then learn that the arytenoid cartilages in the drawing rotate around a point. Since at one pole they each connect to a vocal cord, their rotation either brings the cords closer together or further apart, depending on

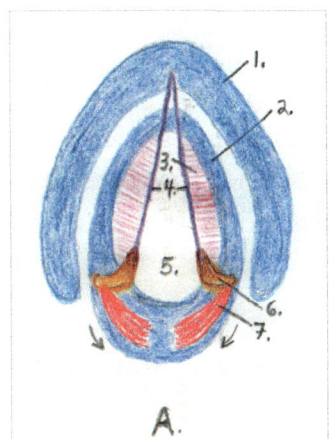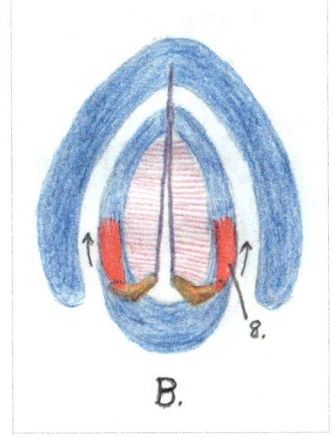

Fig. 1.29 Superior view of the larynx showing muscles that rotate the arytenoid cartilages, thereby causing the vocal cords to move together or separate. **A.** Glottis wide. **B.** Glottis narrow. 1. Shield cartilage. 2. Cricoid cartilage. 3. Conus elasticus. 4. Vocal cords (vocal folds). 5. Glottis. 6. Arytenoid cartilage. 7. Posterior cricoarytenoid muscle. 8. Lateral cricoarytenoid muscle.

Part I. 9th Grade

the direction it takes. OK, so how do they rotate? Muscles, of course, but how are these attached? Again, we figure out approximately where they would need to be located, and indeed, as we thought it must be, we learn about the existence the lateral and posterior cricoarytenoid muscles, which bring the vocal cords closer together or farther apart, thereby modifying the width of the opening between them, the glottis. A side view through a transparent shield cartilage can also help the students visualize what is happening here.

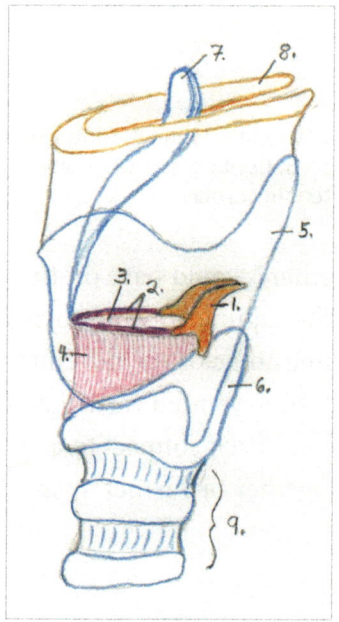

Fig. 1.30 Transparent shield cartilage provides side-view of: 1. Arytenoid cartilages. 2. Vocal cords. 3. Glottis. 4. Conus elasticus. 5. Shield cartilage. 6. Cricoid cartilage. 7. Epiglottis 8. Hyoid bone 9. Trachea.

On this basis, we consider various additional phenomena such as why males have deeper voices, and what happens when the voices of boys suddenly grow deeper; what happens when we whisper, or cough; what is the role of the epiglottis, etc.?

After such, more functional explorations, it is also worthwhile to point out that we actually possess both a wind and a string instrument in service of our voice formation. It is also possible, if time allows, to consider the nature of tone creation in various animals. How, for example, the larynx extends upward into the nasal cavities in various mammals, whereas in birds the larynx (syrinx) is located deep in the chest where the trachea branches. This makes bird song possible, which contrasts greatly with the variety of sounds produced by mammals.[27]

When we expand these considerations from mere voice production into actual speech, we learn that two distinct aspects are involved: phonation and articulation. Having just learned about phonation (voice formation) in connection with the larynx, it now makes sense to continue upward into the head region to find out about articulation, where the vibrating air is shaped in various ways that determine what type of sound is produced. Here again, the teacher can take the students through a series of exercises to help them experience more consciously the unbelievable dexterity involved in the subtle movements of tongue, lips, teeth and palate that modify the shape of the resonating chamber and thus the sound that is created. That the sense of movement is a central factor here will be obvious to the students. What we discover—once again—is an amazing capacity that we take for granted and normally know little about.

It often makes sense to complete this topic area by making the students aware—as we did earlier with sight and hearing—that the involvement of the brain in the activity of speech is not as straightforward and simple as is often imagined.

With the help of a drawing the students learn that there are different areas of the brain (Broca's area, Wernicke's area, etc.) involved and that these are shaped by our own activity. After describing this not easily grasped dynamic, it is often helpful for the students when the teacher dictates a summarizing statement, such as the following from Professor Johannes Rohen:

> *Speech is not automatically organized and guided by the brain's 'speech centers.' Different types of speech centers are located in different places and connected only by complicated, partly hypothetical associative networks. The 'motor' speech center develops only on one side of the brain: on the left in right-handed individuals, and usually on the right in left-handed people. This phenomenon suggests that the development of limb activity, upright walking and hand dominance in the first few years of life provides incisive impulses for the differentiation of both the cerebral cortex and the faculty of speech. In other words, the brain is not the primary initiator of speech.* (2007, p. 214)

---

27 If time is available, students are usually fascinated to learn about the pendulous *laryngeal* sacs that adult orangutan's use as a resonating chamber for their "long call," which can carry for almost a mile.

Part I. 9th Grade

The Human Skeleton

Moving from our considerations of several sense organs and their functions to the second primary topic of this block—the skeleton and muscles—I have found it valuable to start with observations of an entire human skeleton. In other words, we are beginning with the whole before moving to the parts. In this way a context is provided into which all the details that follow can be embedded.

Overview

Most students will have viewed an entire skeleton before. So it makes sense to mention right away that what we are observing is very far removed (abstracted) from real life. To start with, the skeleton is—in contrast to the "real thing"—hanging from a hook because it cannot hold itself upright.[28] And even more obvious is the fact that the bones we are viewing are, in "real life," embedded in muscle and tissues, all of which are enclosed in a layer of skin.

Given all of these limitations, we can nonetheless learn a great deal about the human organization from the observations we are about to make. One way to help bring a bit more life into the skeletal impressions is to have large images of several skeletal drawings by the founder of modern anatomy, Vesalius, available in the classroom. These bring in a wonderful way very human gestures into their representations of the human skeleton.

With a skeleton (or several) unveiled in front of the class, it is often helpful to break the students up into smaller groups, giving them the task of discussing among themselves what they see and then listing characteristics that stand out for them. After gathering the observations made by the groups, it becomes very clear that different principles appear to be at work in the spherical head and in the linear limbs. Whereas the former creates an enclosed space with plate-like bones that are largely immobile in their connections (sutures), the linear ones, by contrast, are very mobile in their linkages and radiate out into the surrounding space rather than enclosing. The students know from experience, too, that the

---

28 This returns us to the theme of gravity, but now we face a not-so-simple causal reality: How does the human being overcome gravity in uprightness? We point to this amazing phenomenon but do not explore it further as it reaches beyond the kind of clear linear causality that the students are at home with at this age.

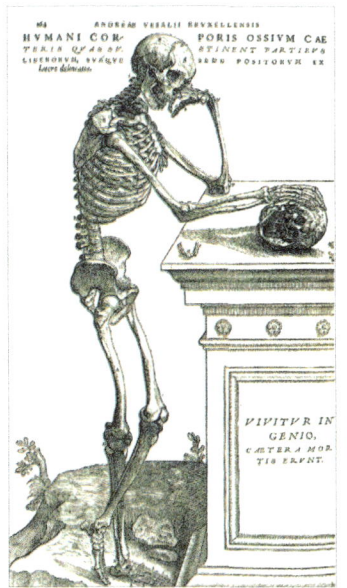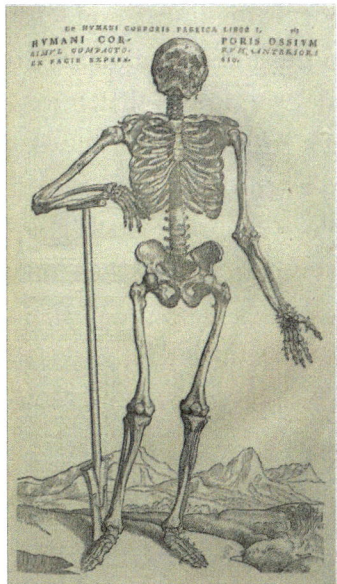

Fig. 1.31 Artistic rendering of human skeleton.
(*Humani corporis ossium cae* by Andreas Vesalius, Wikimedia Commons)

Fig. 1.32 Artistic rendering of human skeleton.
(*Skeleton leaning on pole* by Andreas Vesalius, Wikimedia Commons)

limbs are organs of movement, whereas the head just sits up on top and gets carried around by the limbs and trunk. If we imagine it on its own, we soon realize that the only mobility possible for the head all by itself would be to slide around pushed by flapping jaw movements. But alas, if it pushes too hard and the jaw rolls upward away from the surface, then the head is stuck in that spot forever if nothing else comes along and gives it a shove. The class soon realizes that most of the head community would be easily found in low spots, into which the ever-present force of gravity would eventually cause them to roll!

But, thank goodness, the head has the rest of the body at its beck and call, which carries it up and down hills and anywhere else it desires to go, for the limbs tend to take their guidance from the intentions of the conscious human being. And it is in the head region that we appear to be most wakeful.[29] So, already at first glance, we notice that very different principles are at work in the linear movement pole and the spherical head pole of our bodily organization.

---

29 This is not to say that the head (brain) causes these movements per se. This raises the question of the actual role of the so-called "motoric" nerves, a topic that will not be addressed in this context.

Looking at the linear limbs from a similar perspective, it becomes clear that they too would be immobile and given over to gravity if it were not for a wonderful "teamwork" that informs their activity. Unlike the interlocking, almost frozen relationship of the skull bones to each other, the linear ones are loosely connected through what we call joints—an aspect of the skeleton that we will explore in more detail later. But whence comes the mobility? Clearly something is missing in the skeleton we see before us, and the students, savvy as they are, immediately point out that the muscles (there are about 700 of them) are missing, without which the skeleton we are observing makes no sense at all. These will therefore need, of necessity, to find a place in our considerations in the days ahead.

Having begun with the two extremes of the skeletal structure, we can now include the torso—which lies between them—in this introductory overview.

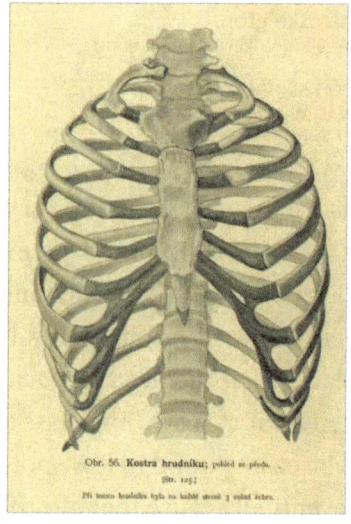

Fig. 1.33 Human rib cage.
(Source: Josef Rejsek [1860–1932], public domain, Wikimedia Commons)

Immediately obvious are the ribs, which individually have a limb-like character, but when taken together, form an interior space reminiscent of the skull above. (The individual ribs are also somewhat flattened, which gives a hint of the plate-like skull bones.) Interestingly, this space is tightly enclosed by the sternum at the head pole, but opens up more and more as we move downward. The ribs remain connected in pairs through costal cartilage for a time, but culminate lastly in the individual and linear "floating" ribs. At both ends of the torso we find the plate-like bones of the pectoral and pelvic girdles. They remind us of skull bones, but are open in their gestures and serve as stabilizing fulcrums for the limb bones. Overall, the students can clearly see how the torso near the head has a head-like tendency, whereas below it is very open and reminiscent of the limbs in its linearity.

After such an overview, in which an overriding organizational pattern has become evident to all, we can move into more detailed considerations of each of the three regions.

The Limb Pole

There are many possible ways to proceed at this point. One that I have found effective begins with the feet and hands. This approach has value in that it leads us in the direction of human uprightness, which, as the students will soon learn, has a decisive influence on many parts of the skeleton. But first, an exercise that gets a conversation going between the hands and the feet. The students are asked to remove shoes and socks.[30] One hand and a foot from the same side of the body are placed close together.

After we make visual observations of similarities and differences between them, we let hand and foot investigate each other in a tactile manner (senses of touch and movement). It becomes evident right away that only the hand has any skill at this. So we proceed with what the hand can discover about the foot and write down our "observations." We then reverse the process and let the foot explore the hand, which, as anticipated, takes place without much success. We

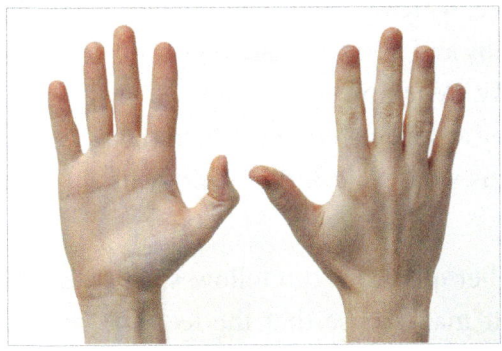

Fig. 1.34 Hands, inside and out.
(Source: Evan-Amos, CC BY-SA 3.0, Wikimedia Commons)

---

30 9th graders often cringe—but enjoy it nonetheless—when the teacher warns them a day in advance to wear clean socks and wash their feet before coming to school the next day in order to avoid possible olfactory embarrassments.

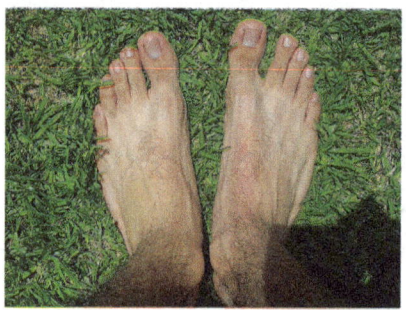 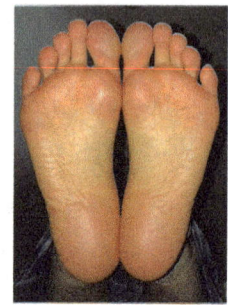

Fig. 1.35 A. Human feet from above.
(Source: Phulvar - Own work, CC BY-SA 3.0, https://Commonswikimedia.org/w/index.php?curid=25046119)

B. Human feet from below.
(Source: Seb2233- Own work, CC BY-SA 3.0, https://Commons wikimedia.org/w/index.php?search=human+soles+of+feet)

then ask: What is it about the hand that makes it so adept at such a task? From that angle we describe many salient characteristics of the hand and record them.

From there it is easy and worthwhile to compare the foot's structure with that of the hand, seeking similarities and differences. Beyond the basic observations that will "jump out" at the students, we can explore numerous other subtleties in the dynamic of grasping and walking if time allows. One important example would be that when walking, running, or jumping, the impact of the body weight puts a tremendous amount of pressure and force on the foot. When a person is running, the force applied to each foot when it contacts the ground can be as much as 2.5 times their body weight. The bones, joints, ligaments, and muscles of the foot absorb this force, which greatly reduces the amount of shock that is passed on to the lower limbs and body. The arches of the foot play a key role in this shock-absorbing ability. When pressure is applied to the foot, the arches will flatten somewhat, thereby absorbing energy. When the weight is removed, the arch rebounds and gives us what we call the "spring" in our step (Biga, Dawson, et al. 2019).[31]

Building on such observations, other questions can follow, such as: Based on their skeletal characteristics, does it make sense that the feet are "on the ground" and connected to the legs, while the hands are above and at the tips

---

31 An excellent resource for additional perspectives in this direction can be found in the article "The Feet Reveal the Will" by Norbert Glas (2004). For more anatomical/structural perspectives, see J. Rohen (2007, pp. 96–98).

The Human Skeleton

(not the base!) of the arms? Silly question it might seem, but interesting to make a case for why things are the way they are. After discussing such things we can go one step further and contemplate what our life would be like if we were to lose our feet, and how would it change, in a different way, if we were to lose our hands.

Such considerations can lead to clear insights and a strong feeling for a basic structural pattern (design, plan) that hand and foot share, one that can be modified in very different directions depending on the field of forces that they are involved with. (In our case, it is the difference between bearing the weight of an entire body [feet], and being freed from that task by uprightness [hands].) Such deliberations regarding basic organizational patterns—where the proportions of the parts vary greatly depending on where they are in an organism and what they do there—provides a small, but first taste of very important ideas (bauplan, metamorphosis, etc.) that will receive a good deal of attention in the 11th and 12th grade biology blocks.

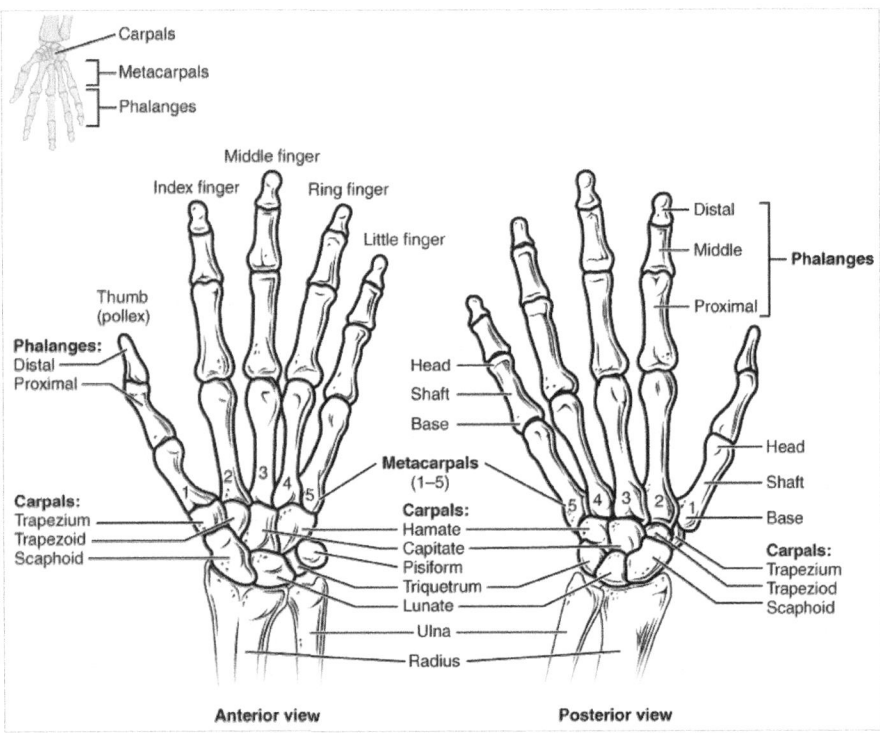

Fig. 1.36 Bones of the human hand.
(Source: OpenStax College, CC BY 3.0, Wikimedia Commons)

Part I. 9th Grade

From here, we can take the comparative approach we have been practicing and widen it to include the arms and legs. Both are clearly made up of three sections that are connected by joints. In each case the number of bones per section increases as we move from the center toward the periphery. In the case of the arms, we move from one upper arm bone (humerus) to two forearm bones (ulna, radius), and then to three upper and four lower carpal bones, followed lastly by five metacarpals and five fingers (each of which—with the exception of the thumb—contains three phalanges).

Looking at the legs, we find the same pattern except for the tarsal bones, which are more differentiated and overlap to form a strong arch.

In light of the relationship between above and below relative to the forces of gravity, it is not surprising to the students that the increasing number of bones toward the periphery of the arms enables a wide range of flexible movements and possibilities for self expression (gestures), while in the legs the same pattern is much more oriented toward stable support for the whole body, as well as locomotion. This is particularly evident when comparing the rotational capacities of the radius and ulna with the lack thereof in the tibia and fibula. The

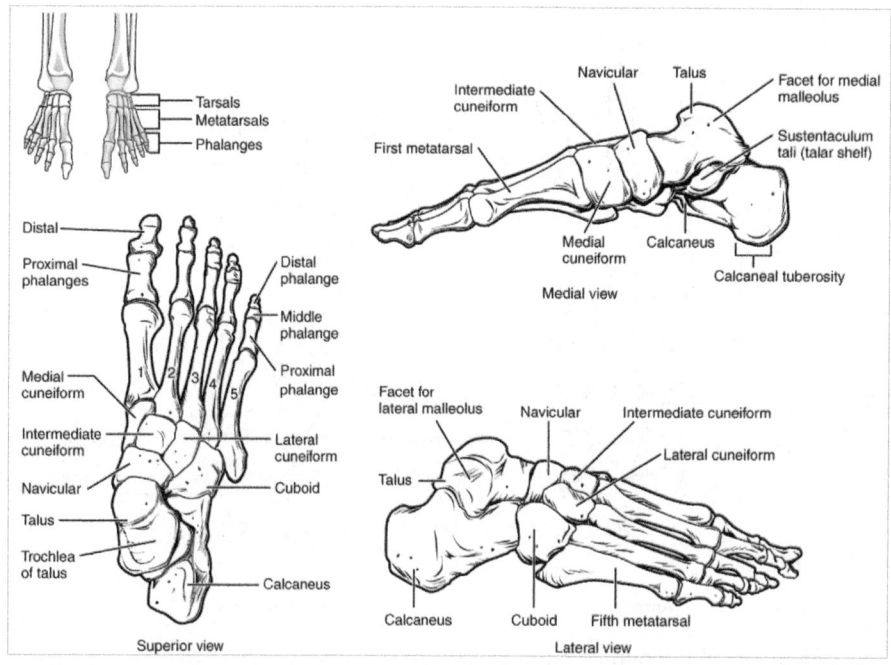

Fig. 1.37 Bones of the human foot.
(Source: OpenStax College, CC BY 3.0, Wikimedia Commons)

rotational possibilities of radius and ulna are integrated into the flexibility found in the wrist (carpals), and thereby in the grasping dynamic of the hand. The lower leg, by contrast, does not rotate. Its range of motion is very reduced in the service of stability. Its bones (and tendons) are larger and stronger. The twisting flexibility we find in the wrist bones has been largely reduced in the ankle to a 90-degree bend.

We can extend such comparative explorations yet further by asking how the pectoral girdle above and its counterpart, the pelvic girdle below, appear in relation to one another.

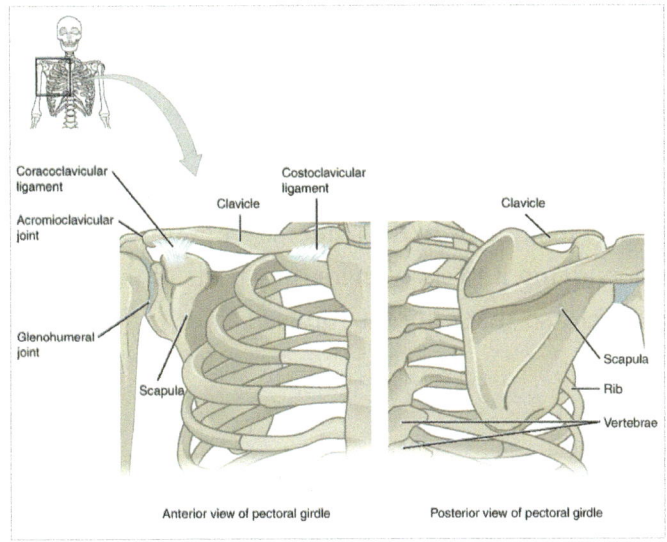

Fig. 1.38 Pectoral Girdle.
(Source: OpenStax College, cnx.org, Creative Commons BY 3.0)

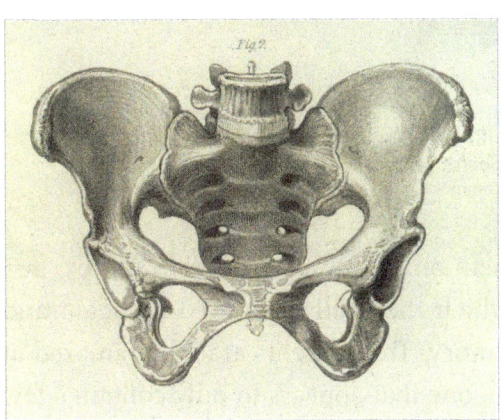

Fig. 1.39 Drawing of the female pelvis.
(Source: Wellcome_L0038287.jpg, CC BY 4.0, Wikimedia Commons)

Part I. 9th Grade

Not surprisingly, the shoulder girdle is much freer in its connection to the arms compared to the stabilizing support provided by the pelvis for the legs. Shoulder and hip joints also provide an interesting contrast, as do the heads of the humerus and femur bones. Additional contrasts between above and below can be observed if time allows, but one in particular fascinates the students and gives rise to reflection: the fact that the primary arm joint (elbow) bends forward, while the primary leg joint (knee) bends backward! How would it be if this were the other way around? An interesting assignment for the next day is to have them observe how these joints bend in the four-legged animals they are familiar with.

The Head Pole

Although we have already begun to speak about the torso by observing the pectoral and pelvic girdles, I usually move on at this point to the counter-pole of the limbs, the skull, before rounding off with the trunk that lies between them. Because the skull is so familiar to the students—but only in a superficial way—it is helpful when studying it to show them only parts of the skull at a time. It can be held, for example, so that only the top is visible (with the rest enclosed in a cloth).

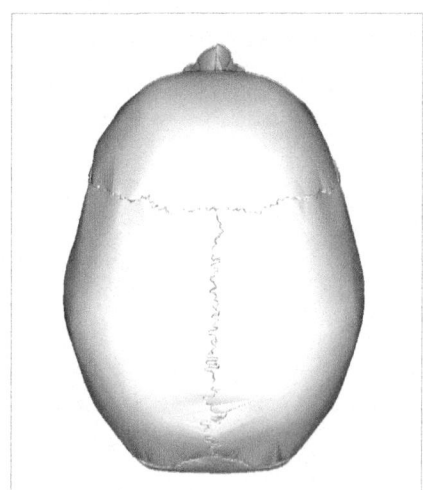

Fig. 1.40 Human skull from above.
(Source: Anatomography, CC BY-SA 2.1 JP, Wikimedia Commons)

I have found it effective to give them only a "quick glimpse" (approx. 3–5 seconds) the first time around, after which the skull is covered up again and first impressions are gathered from memory. The students are often amazed at the uniform oval surface they have seen, one that appears to only contain a few

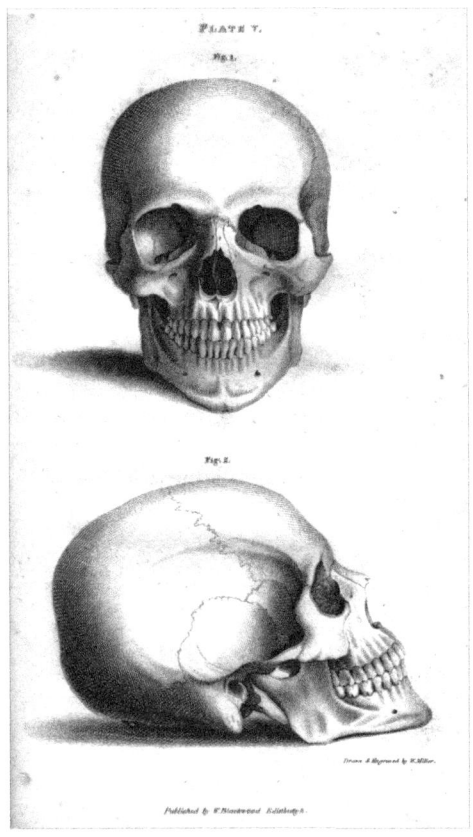

Fig. 1.41 Frontal and lateral views of human skull.
(Source: William Miller, public domain, Wikimedia Commons)

squiggly lines across its surface. A next viewing from this one-sided perspective can then take longer and provide the opportunity for more detailed visual observations, touching, etc.

The next "quick glimpse" can be of the facial region. What a different world! Again, we gather first impressions from memory after a quick glimpse, followed by a second longer look, after which descriptions become quite detailed.[32]

---

32 I find it very useful, when the students are observing phenomena that they are familiar with, to ask them not to use the names of things that they already know, but instead to describe what they see in terms of form, color, size, etc. Instead of saying teeth, for example, they would describe two rows of tiny white "nuggets" that appear to be inserted on one of their ends into the larger, less white surface above, or below, and so on. Freeing them from their familiar vocabulary and categories leads to much more conscientious and detailed observations.

After precise observation and gathering, we step back, compare these two "worlds," and reflect on how such forms reveal something of the activities that take place there. A side view can then follow, which contains parts of both poles and shows how they merge into one another.

Since we don't have another organ with a similar "organizational plan" for comparison (as we did with hand and foot), it is worthwhile to compare the human skull with a mammal's. This, as in all comparisons, allows us to highlight certain prominent characteristics through the way they differ from those of the comparison-partner. I have often used a cow skull for this purpose, because it highlights vividly the size and spherical nature of the human cranium on the one hand, the reduced jaw and the vertical nature of the facial skull (as seen in profile) on the other.[33]

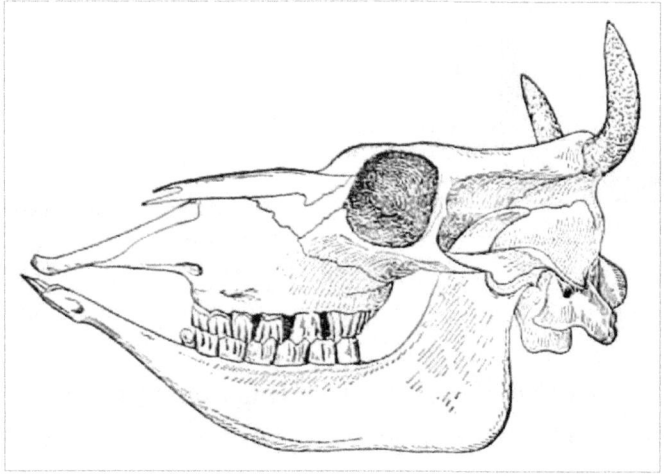

Fig. 1.42 Cow Skull.
(Source: NIE_1905_Cattle_-_cow_skull.jpg)

---

33 I have found it fruitful to introduce the cow skull using a similar method to the one described above for the human skull—the so-called "quick glance, first impression" approach. In this case the emphasis would be on a side (lateral) view of the entire skull, since the pivotal characteristics for this comparison are most evident from that angle. The students will, of course, want to look at the skull from various angles, and this can happen once the original and pivotal "first glance" exercise has taken place. Additional comparisons with other skulls (rodents, carnivores, birds, etc.) would certainly be possible and interesting, but I lean toward reserving this for the 12th grade zoology block, where such comparisons can be embedded in a much more expansive inquiry into various animal types and how they relate to the human being.

After highlighting the polarity between the facial and cranial aspects of the skull, we move on to various other features such as the teeth,[34] the jaw as the only (visible) moving part in the head, the location of the foramen magnum in humans in contrast to the cow, etc.

Leaving the skull for the time being, we now take another, closer look at what lies between the head and limb poles of the body—the torso.

The Torso

In addition to the formative tendencies in the torso that we noticed at the beginning of this chapter, we now look a bit more closely at the how the trunk—in contrast to head and limbs—is composed of repetitive segments (metamerism). We noted earlier how the twelve rib pairs gradually change from an enclosing gesture near the head to an open one as we move downward.[35]

Taking a closer look at the vertebral column, we can find several aspects that "ring true" to the thinking of a 9th grader. First of all, the vertebrae have a basic structure that changes gradually from the lumbar vertebrae below to the cervical vertebrae near the head. The way they change makes sense to the students: Below, in the load-carrying lumbar area, the vertebral body is large, and, as we move upward along the spinal column, this body becomes smaller and smaller until it is missing altogether in the uppermost vertebra, the atlas. In its place, the axis below it has developed a vertical "tooth" (dens) around which the atlas—and with it the head—can turn.

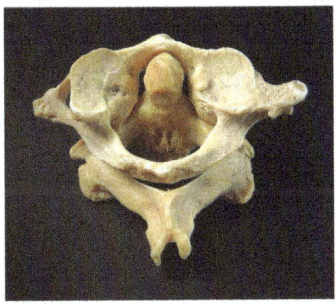

Fig. 1.43 Cervical vertebra (Atlas C1) and lumbar vertebra (L3) for comparison.
(Source: Anatomography, CC BY-SA 2.1 JP Wikimedia Commons)

---

34 This is also an area that can be compared with various mammals in greater detail in the 12th grade.
35 The intercostals muscles that connect the ribs will be considered in the 10th grade in connection with the lungs and breathing.

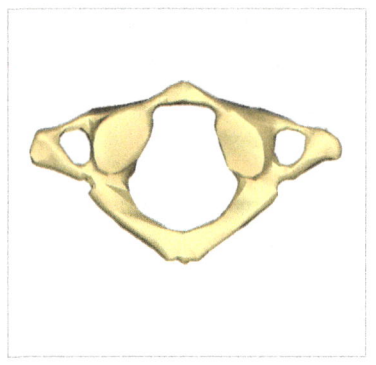 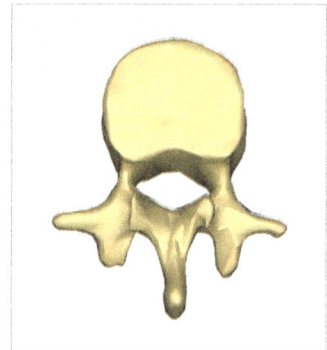

Fig. 1.44 Atlas and Axis.
(Source: MAKY.OREL, CC0, Wikimedia Commons)

Moving downward from the lumbar region we find, on the other hand, the five vertebrae of the sacrum fused together, thus eliminating any movement at all between them—which is the complete opposite of the atlas/axis relationship. At the very bottom, the approximately four coccygeal vertebrae meld to form the so-called tailbone.

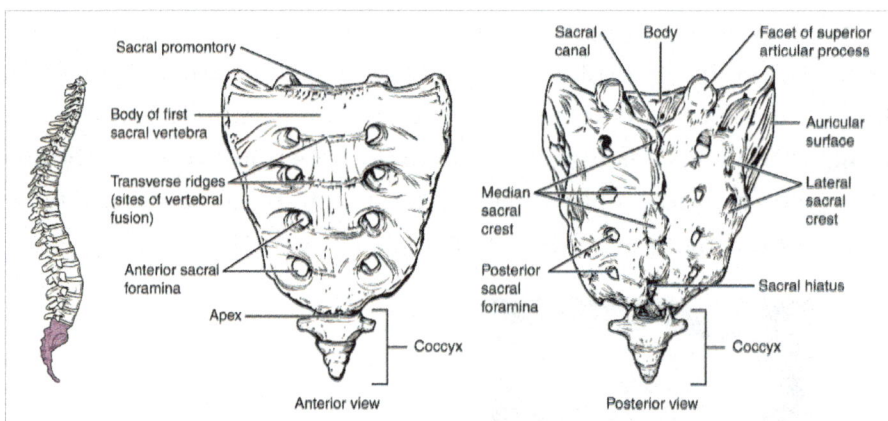

Fig. 1.45 Sacrum and coccyx. (Wikimedia Commons)

It is familiar to those who have fallen on it and experienced the significant pain that it can cause (often a long-lasting one!). The students empathize—and laugh—when I tell them how my younger sister once kicked me in the tailbone and it hurt for months thereafter. (Just what she wished for! She later became a nurse and—I tell the students—the theory I propagated in those days was that she was trying to make up for all the pain she had caused me by becoming a nurse and helping minimize the pain for others.)

If we compare our ability to move our neck and head with the mobility of our thorax (upper back) and lumbar (lower back) regions, we find a major difference. The head can tip forward and backward and it can rotate (thanks to the atlas and axis). In order to see how much mobility they have in the other two areas we stand up and do various exercises to assess this. The teacher can lead as we see how the three regions differ when we bend forward and backward, right and left. We lie down and gauge how well our chest can do situps, for example. We can even explore which regions are moving the most (besides the limbs) in various contemporary dances. (For each dance one or two students can come up at a time and demonstrate while the others observe.) A hula hoop is also fun to try out. In sum, we are most restricted in the chest and upper back region in our range of movements (even though, on a finer scale, they are constantly in motion through our breathing). The students also enjoy imagining what it would be like if the mobility of the three areas were changed around. If, for example, we could bend at the chest like we can do with the lower back, while, at the same time, the lower back became as immobile as the chest, etc.

Next we ask: How would it be if our spine did not have this flexibility, but were merely a solid rod made of bone? Again, we stand up and try to maneuver with a completely rigid backbone. In this way we experience the significance of the metameric nature of the spinal column. We consider the curvature of the column when we look at it from the side.

The double-S shape is easy to see. It has, namely, four prominent curves: two convex curves (the neck lordose and the lumbar lordose) and two concave curves (the chest kyphose and the sacral kyphose). After noting this, we consider what significance such a structural form might have. Students jump from their chairs (or something similar) and notice the spring-like effect the double curvature has—it absorbs much of the shock. A student carrying a heavy backpack reports a similar phenomenon. The flexible curves inflect more to absorb some of the weight and then rebound ("spring back") when the pack is removed. In terms of flexibility, shock absorption and weight accommodation, what would it be like if we had just one long curve? We try walking with our spines curved in one long arc toward the front, and then the reverse, with it curving toward the back. We discuss the challenges (and positives?) each would create.

But how are these curves and the variations in flexibility that we have been observing even possible if bone sits on bone like it appears to do in the skeleton?

Part I. 9th Grade

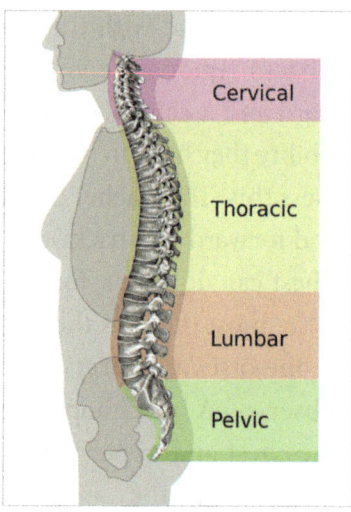

Fig. 1.46 Human vertebral column, dorsal and side view.
(Source: Creative Commons Attribution-Share Alike 4.0 International))

Many of the students know, of course, that something flexible is located between the segments—the intervertebral disks. These discs contain an inner gelatinous pulp that is elastic and compressible. The pulp is surrounded by a strong "collar" (annulus fibrosus) that limits the expansion of the pulp within when the spine is compressed.[36] The discs are thickest in the cervical and lumbar regions, which helps give them more flexibility. Since they broaden and flatten out some each day, we have the opportunity to the gather some experimental data. We ask the students to measure how tall they are before going to bed that evening, and then again soon after they get up in the morning. They will usually gain 1–3 cm. in height overnight! (The discs themselves actually make up about 25% of the length of the spinal column.)

Uprightness

After the extensive comparisons we have made between the limbs, torso and head, it is important for the students to learn that a number of the characteristics we have observed are not there in young children. In many instances, the forms we are familiar with develop only when the child raises its body into the upright and engages the field of gravity in an active way. This engagement calls forth a reshaping of key aspects of the skeletal structure.

---

[36] Some students will have heard of slipped disks (hernias) which can be explained to them in this context.

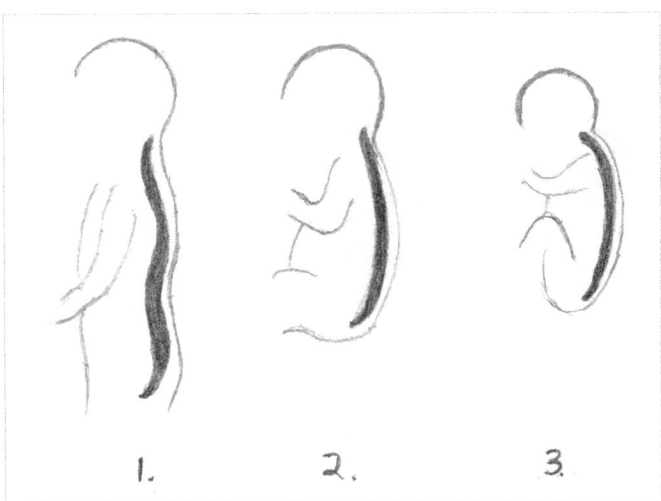

Fig. 1.47 Development of spine's "double S" curvature in connection with uprightness. 1. Six years. 2. Nine months. 3. Birth.

For example, at birth, the double-S curvature of the spine has not yet developed. Although flexible, the infant's spine has very little curvature. Moreover, the individual vertebrae are all roughly the same size at this age!

Once we make the students aware of this, they discuss what could bring about such significant changes. Thinking chronologically, we note how the child gradually begins to deal with gravity through its movements. When it begins to raise its head at about four months (which requires working against gravity in the head/neck, and, to some degree, shoulder regions), the neck curve (cervical lordose) begins to form and the neck muscles grow stronger. At approximately nine months, when the baby can sit and carry its head freely, the neck curve is complete!

Through the struggle with gravity while learning to sit, the curvature of the thorax (thoracic kyphose) becomes more pronounced, and when the child begins to stand, the curvature of the lumbar region (lumbar lordose) forms. At the same time, the vertebrae in this area—which now must bear much more weight than hitherto—become larger. By the age of six, all of the spinal curvatures are complete. In the years that follow, the sacral vertebrae will begin to fuse, and later yet, the coccygeal vertebrae merge.

A striking transformation also takes place in the feet. The newborn child's feet are relatively wide and flat.

Part I. 9th Grade

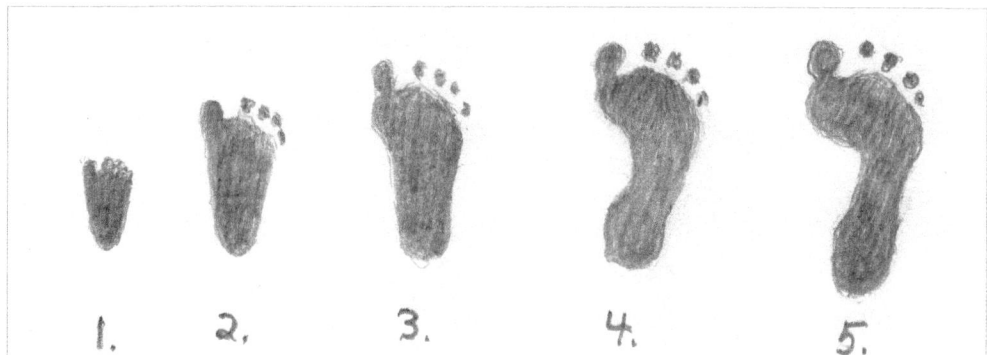

Fig. 1.48 Development of the foot's longitudinal arch in childhood. 1. One month. 2. Two years, one month. 3. Three years, eleven months. 5. Five years, eight months. 5. Six years, nine months. (After Holdrege 1996)

As the footprints above show, the characteristic longitudinal arch of the adult foot develops as the child rises into the vertical and begins to overcome gravity through its own will. If gravity alone were to dominate, then the longitudinal and the transverse arches such as seen below would not exist.

Fig. 1.49 The transverse arch of an adult human foot. (After Benninghoff & Goerttler 1975)

With the help of drawings and handouts we can see how the midline of the heal bone (calcaneus) rotates from a partially horizontal position in infancy into uprightness (thereby counteracting gravity) when the child begins to stand.

The complex interworking of the numerous components involved in forming the foot as the child begins to stand and walk is well summarized for the students in the following quotation:

> *There is no other part of the human body where you can see so clearly the harmonization and coordination of a functional system of bones, cartilage, muscle and tendons as in the foot. Every part carries and supports the others; every disturbance of this balance—for example the*

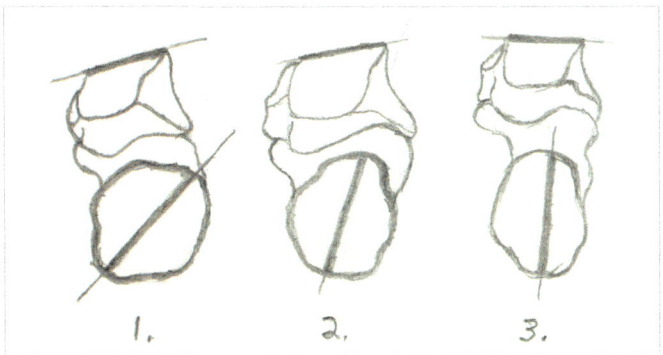

Fig. 1.50 Rotation of the heel bone into uprightness.
1. At birth. 2. Age two. 3. Adult. (After Benninghoff & Goerttler 1975)

weakness of a muscle—leads to changes in position, and results in the deformation of the whole foot. (Straubesand, in Kranich 2003)

At birth, the legs are surprisingly short. They make up about one-third of the entire body length, in contrast to the adult, where, on average, they comprise over half the body length. To reach these proportions, the legs must grow more than any other organ in the human body—and this growth takes place against the increasing weight of the body above it!

In the infant, they bow to the outside and a straight knee joint has not yet formed. This means that when the child first stands, gravity has a strong influence. But through growth and the straightening of the knee, this influence becomes less and less as the typical human leg form develops.

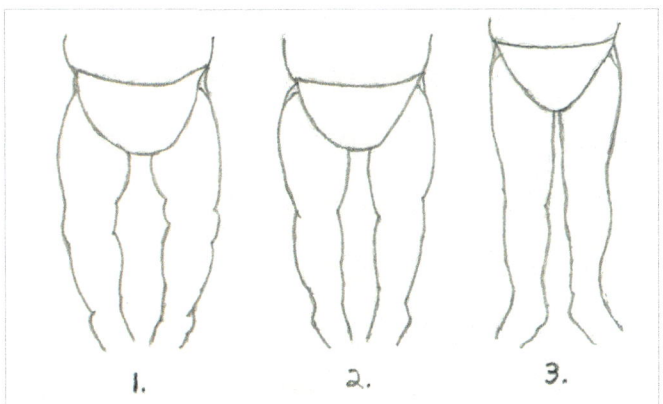

Fig. 1.51 Straightening of human legs from infant to toddler.
1. At birth. 2. Six months. 3. Two years.

Part I. 9th Grade

Through transformations such as those just described, the students develop a sense for the central role that uprightness plays in human existence. They learn to see how human activity—in this case the striving to stand up and walk—has a formative influence on the body. (This is a topic we will take up again in the 10th grade.)

Such changes also speak to the contrasts we have already found in the lower and upper extremities, both of which express clearly the difference between the weight bearing task of the lower limbs and the emancipated freedom given to the upper ones through uprightness.

At the conclusion of such considerations, I often use a quotation by Erwin Strauss that describes the dynamic quality of human uprightness.

> *Upright posture characterizes the human species. Nevertheless, each individual has to struggle in order to make it really his own. Man has to become what he is. ... While the heart continues to beat from its fetal beginning to death without our active intervention and while breathing neither demands nor tolerates our voluntary interference beyond narrow limits, upright posture remains a task throughout our lives... In getting up, in reaching the upright posture, man must oppose the forces of gravity. It seems to be his nature to oppose nature in its impersonal, fundamental aspects with natural means. However, gravity is never fully overcome; upright posture always maintains its character of counteraction. It calls for our activity and attention. (1966, p. 141)*

Joints

After portraying the wonder of uprightness, we can return to more transparent—easy to grasp—causality that is evident in the joints, which dynamically link the skeletal elements to each other and to the muscles that move them. Indeed, it is important to emphasize at this juncture that the skeletal structures we have been considering can only be understood in the context of their movement potential. This movement is only possible because the various bony components are linked to each other by joints.

A good starting point is to inspect an upright skeleton from top to bottom (or in some other sequence) and look for mobile connections between the bones. In doing so, we discover differing degrees of mobility. In the head region we find—not surprisingly—only one freely-movable joint: where the lower jaw connects to the base of the skull in the temporomandibular joints.[37] Although not obvious in a wired-together skeleton, it is one of the most mobile joints in the human body. What would be considered a dislocation anywhere else is normal for the jaw. The head of the mandible slides out of its socket easily when the mouth is opened wide, and can move back-and-forth, as well as side-by-side! This flexibility occurs only in humans and helps make speech possible.[38] Moving downward we come to the torso and the joints that connect it to the limbs. These joints (shoulder and hip) are more mobile than the joints found peripherally in the limbs. Whereas shoulder and hip joints allow movement in all three planes of space (tri-axial), the elbow joint can move in only two dimensions (rotation and extension/flexion); the wrist is also limited to two dimensions, while the middle and terminal joints of the fingers are reduced to a simple plane. When you move down the leg, you see a similar pattern, although—not surprisingly based on what we have learned so far in this block—the leg joints tend to be less mobile than those in the arms (Rohen 2007).

After these observations, the students are usually interested in learning how joints function in general and how the variation in mobility just observed

---

37 Sutures are also joints, but ones that are highly constricted in their range of movement.
38 The right angle of the jaw to the base of the skull (as well as the inhibition of jaw growth) that makes this possible is one of the most significant skeletal features that distinguishes *Homo sapiens* from early hominids and apes.

Part I. 9th Grade

comes about. This being the 9th grade, I try to approach these questions in a way that reveals straightforward causality.

To start with, it is helpful to note that bones can connect in several different ways. As already noted, the sutures in the skull are very limited in their movement. These joints are joined together by a thin layer of fibrous connective tissue. Then there are the slightly more movable joints of the vertebrae that are linked together—as we saw earlier—by the malleable invertebrate discs. What we normally call joints—the kind we just observed during our inspection of the skeleton—are synovial joints. All the joints found in the limbs fall into this category.

So what are the characteristics of a synovial joint? Rather than describing abstractly a generic synovial joint, I recommend choosing one specific joint and exploring it in detail. The knee joint, for example—as the largest and most complex joint in the body—lends itself to this purpose.[39] If we manage to investigate the structure and physiology of the knee in a clear and causally transparent fashion, the students will feel they have developed some expertise in an area of the human body that is very relevant to most individuals who engage in athletic activities—and to all of us who have grown older. In the current context, I will not do this, but only refer to a few key aspects of synovial joints, going into less detail than I do with the 9th graders in class.

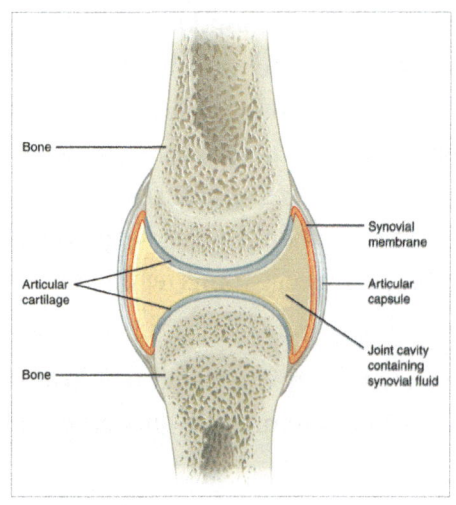

Fig. 1.52 Generic image of a synovial joint.
(Source: OpenStax College, CC BY 3.0, Wikimedia Commons)

---

39   Although it is actually three joints in one!

With the help of a drawing, we describe a fluid-filled joint-cavity that separates the articulating bones and permits substantial freedom of movement between them. That such joints require the support of band-like ligaments makes sense to the students, as does the lubricating function of the synovial fluid that reduces friction between the cartilage-covered bone surfaces that meet in the capsule. They also understand how its viscosity helps to absorb shocks.

At this point, the students usually like to hear how—with their new understanding of joints—a number of curious phenomena that they have met in their own lives, or have heard about, can be explained. Such phenomena might include: Why do joints sometimes swell up? What takes place when one sprains an ankle? Why do knuckles "crack"? What is the key to being "double-jointed"? Why is warming-up before exercise wise? What is a torn meniscus? What is arthroscopic surgery? What is a knee replacement, etc? There are, it turns out, many interesting questions that come alive among the students once they begin to learn about the structure and functioning of human joints (Marieb & Hoehn 2012, Tortoro & Derrickson 2013).

After such considerations we can raise a new question for discussion. If the same basic organizational principles can be found in all synovial joints, then why are they so different in terms of the freedom of movement available to them? We compare, for example, how we can rotate our arms at the shoulder along its long axis, but we can't do this with the thumb, despite its great mobility. We also look at how we can rotate our lower arm down below as a whole, but not at the elbow, and not at the wrist, etc. ... These are things we do together in class, as we try to awaken questions in the students about the nature of the joints. All of this should happen before we "solve the riddle" by learning about the different types of joints.

Looking back to the skeleton again, it is clear that the forms of the bones that meet in the various joints differ considerably. So how does form influence function in these instances? At this point I usually give an overview of the primary types of synovial joints, with simplified drawings that indicate planes of movement, followed by short explanations and an exploration of the skeleton to figure out where examples of each type might be found.

In the shoulders and hips we find **ball-and-socket joints,** where the ball-shaped head of one bone fits into the cup-like socket of another. These joints can rotate in all directions (multiaxial).

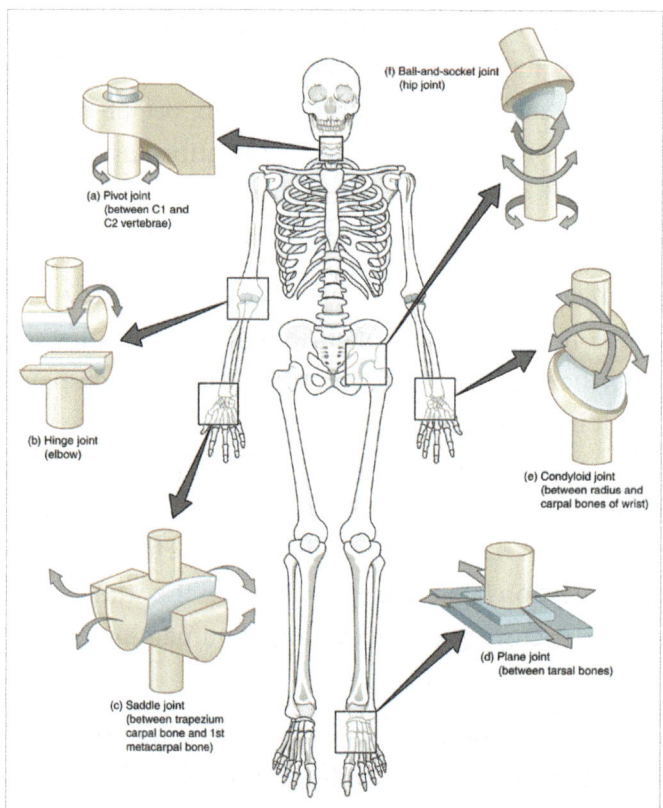

Fig. 1.53 Schematic illustration of the primary types of synovial joints indicating their planes of movement.
(Source: OpenStax College, CC BY 3.0, Wikimedia Commons)

Saddle joints occur where each bone is saddle-shaped and the way they fit together allows them to move in different planes, but not to rotate. An example of this is the carpometacarpel joint of the thumb.

When the oval articular surface of one bone fits into a complementary depression in another, we speak of a condyloid joint. The wrist and the knuckles between the metacarpels and the phalanges exemplify this. They can move in different planes, but cannot rotate.

Pivot joints can only rotate. Here the cylinder-shaped end of one bone protrudes into a sleeve or ring-shaped cavity of another. Here the only movement possible is the uniaxial rotation of one bone around its long axis. Examples of this are the joints between the radius and the ulna, and between the atlas and the axis at the top of the vertebral column.

In **hinge joints** the cylindrical convex surface of one bone fits into the trough-like concave surface of the other. Motion is only possible in one plane (uniaxial), bending or straightening. The elbow and the interphallangial joints exemplify this.

In **plane joints** (gliding joints) the surfaces are flat or slightly curved and allow only short gliding movements. These can be found in the intercarpal and intertarsal joints, as well as the articular processes of the vertebrae.

The drawings of the joints shown here appear very mechanical, but I do this for the sake of demonstrating clear, straightforward causality. Although organisms are much more complex and reveal much higher levels of interactive causality, it is not false to make it appear so clear-cut when we focus on certain limited aspects of the body, such as the joints. To do this all the time and in reference to many parts of the organism would be false and would destroy the sense for how an organism differs from a mechanism. Here we do it in limited areas and with a specific age group. Things will need to grow more complex and interwoven in the higher grades.

Part I. 9th Grade

Muscular System

After all this work on the structural make-up of the human being, it is finally time to get things moving. Since the human body has only about 700 individual, voluntarily-controlled muscles (Totoro & Derrickson 2013, p. 337), it is not very easy to give a clear picture of them all! So, as we now set out to investigate the "mover and shakers" of the bones and joints, our primary point of view will be—once again—which aspects of this topic can provide "nutrition" for the birthing of the 9th grader's new cognitive capacities? How can we capture 9th-grade-suitable aspects of this huge topic without getting lost in all the details?

My experience has been that two levels of consideration—at least—speak to students of this age: first, the way that our bones work as levers and our joints as fulcrums when we bring about movement;[40] second, the different levels of muscular activity from the unconscious (involuntary) activity of our smooth muscle tissue to the muscles of facial expression in the head that reveal much of our inner life. (The latter provide a small taste of what will be a central aspect of the 10th grade biology block.) Many other aspects of the skeletal muscles can be considered—the choices are endless—and will depend on the class one is teaching, the enthusiasm the teacher might have for this or that aspect of the muscular system, and—as always—the time available. In this text I will limit myself to the two aspects just mentioned.[41]

Leverage

Since the skeletal muscle fibers consist of long contractile fibrils (myofibrils), they can contract in only one direction, which necessitates two antagonistic muscle groups that pull in opposite directions around each joint. Each time a "flexor" contracts, the corresponding "extensor" must relax.[42]

---

40   As mentioned earlier, it is important to confer with the 8th grade teacher to see what they have or have not done in this area the previous year.
41   It should be remembered that the smooth, involuntary muscles receive some reference in the 10th grade block, for example in the blood vessels and gastro-intestinal tract. The cardiac muscle, as well as striated muscles such as the intercostals and the diaphragm, also come under consideration in the 10th grade.
42   Muscles can also perform a static strengthening function when, like elastic belts, they pull on bones in various areas of the body. They function thereby as a kind of bracing system that relieves stress and allows our skeletal bones to be less massive than they would otherwise need to be (Rohen 2007).

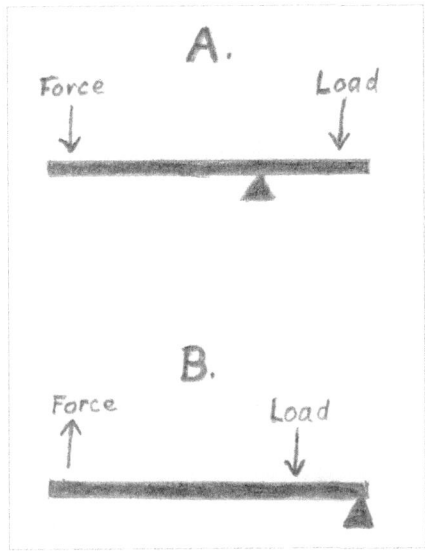

Fig. 1.54 A. First-class lever.
B. Second-class lever.

When they produce movement, bones act like levers with the joints serving as fulcrums. If the students no longer have a clear sense for the workings of levers from a grade school mechanics block, a few quick explanations can refresh their memories.[43] However, in contrast to the familiar first-class levers (child and adult on a seesaw; car jack, scissors) and second-class levers (wheel barrow), a third-class lever is what we usually find in the human being.

In my experience, it usually helps to clarify the basic aspects of a lever and leverage using familiar instances of a first-class lever. Using a car jack as a example, we let the students analyze what is going on. It quickly becomes clear that the tremendous force needed to raise the car must come from the much greater length of the "Force Arm (FA)" relative to the "Load Arm (LA)" and the much longer distance that it moves. We conclude: When the load is located close to the Fulcrum and the Force that is applied far from it, then a small Force exerted over this distance is able to move a large Load over a short distance. This is called a power lever. With lots of movement and relatively little force, we can move a large load small distances.

---

43   It's often nice—and effective—to see if there are students in the class who can explain this to the others.

Part I. 9th Grade

What happens if we reverse the situation? Here we would be at a mechanical disadvantage, requiring more Force to be exerted than the weight of the Load itself. On the other hand, the Load will move rapidly and a much greater distance than the Force Arm. This is known as a "speed lever." As an example we show how an object can be jettisoned into the air with such a lever. We sum up all of this with an equation they may already know.

Load x Load Arm length   =   Force x Force Arm length

(L x LA   =   F x FA)

With this preparation behind us, the fun can begin. For now the question becomes: How does this apply to our limbs? Starting with the arm as an example, we see one big difference to the car jack right away. We do not have one continuous lever arm that pivots around a joint (fulcrum). So "what is what" in this constellation of factors? Where is the Load Arm, the Force Arm? The elbow joint is clearly the Fulcrum, but what else fits into our categories above? It does appear that the Load Arm would run from the elbow to the hand, but where is the Force Arm? Reflecting further, we note that the force in this context must be applied by the biceps muscle onto the Load Arm near the elbow. Levers that are organized in this way are known as third-class levers. A quick diagram is helpful here.

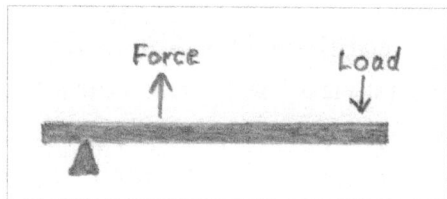

Fig. 1.55 Third-class lever.

Since we cannot move the location in which our biceps inserts into the lower arm, it may help to use a different example of a third-class lever like a shovel (or a broom, a fishing pole, tweezers) to test what happens when the location of the applied Force is varied. If we call the hand we place at the end of the handle the "Fulcrum Hand," and the hand that lifts the Load our "Force Hand," we know the latter can find itself in various locations along the handle between the Fulcrum and the load on the blade. We can now experiment using different "Force Hand" locations (which corresponds to shifts in the connection point of the biceps on the forearm (Ulna) and notice how things change. For

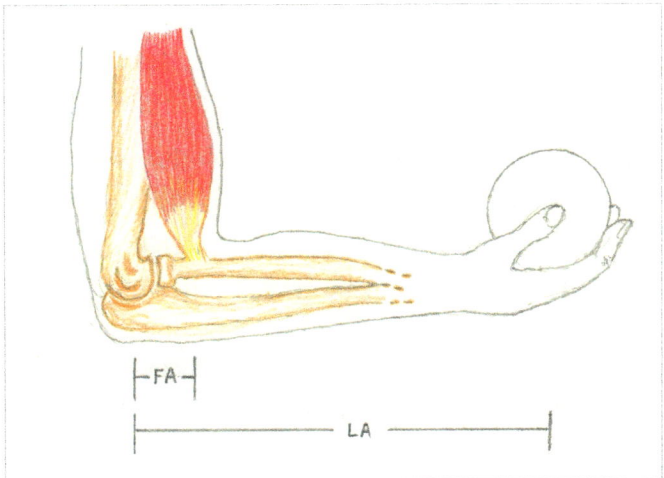

Fig. 1.56 Biceps muscle as part of a third-class lever with a very short Force Arm relative to the Load Arm.

one, the closer I move my "Force Hand" toward the blade, the lighter the load gets, the closer it is to the Fulcrum, the heavier it becomes. But something else changes as well. If I want to toss (pitch) the load from the blade a long distance (assuming it is not too heavy) where do I best place my "Force Hand"? Near the blade I can generate very little speed, closer to the Fulcrum the speed increases greatly and I can toss it much further. It is now clear: The distance between the elbow and the connection point of the biceps onto the forearm represents the length of the Force Arm.

After these experiments, we look at how our arm is constructed again and realize that the insertion point of the bicep muscle so near to the elbow has major consequences. Such a short Force Arm affects the quantities in our leverage formula significantly. We are at a serious mechanical disadvantage. In some classes it is worthwhile to solidify this insight by "doing the math." A few quick measurements will allow us to enter approximate data into the equation. A typical forearm would be around 12 inches in length, and insertion point of the bicep 2 inches from the elbow joint. If we apply this to a five-pound load we get the following result:

5 lbs x 12" = ? x 2" which means → 60 lbs = 30 lbs Force x 2". In other words, in order to lift a 5 lb. object at the end of our 12" long forearm, we have to create 30 lbs of force! We let one of the students hold a heavy bag of books using one arm with the elbow at a 90 degree angle (as in the drawing above). Soon

exhausted, they are then allowed to let the bag hang downward with the arm extended. Much better! Why? The length of the Load Arm is now essentially zero, leaving only the weight of the load to be carried (Von Mackensen et al., 2004).

Noticing that the connection points for the other levers of our body must also be close to the joint, we conclude: Our limbs are built in such a way that we are at a serious mechanical disadvantage! We need much more force to act in the world around us than would be the case if our muscles were inserted farther from the joints. That is so unfair! But, slow down, could there an upside to this arrangement? Yes, indeed! As we saw earlier, in a lever there is an inverse relationship between mechanical advantage and the speed + span of a movement. Therefore, the insertion of the muscle so close to the joint allows us to create much more speed in our movements and gives us a much larger range of motion. Elegance and fluidity of movement are made possible thereby.[44] John Travolta and Olivia Newton-John (younger teachers may need to Google them or think of a more contemporary example) would never have been able to dance the way they did in *Grease* if their muscles had been inserted farther from their joints!

Different Levels of Muscular Activity

In order to bring a more qualitative aspect to our study of the muscles—one that goes beyond what only speaks to the "quantitatively-oriented" mind—I have found it worthwhile to give a quick overview of various types of muscles in the human body that serve us in quite different ways.

But first a more general observation. As one muscle expert, Simon Pressel (2004) suggests the following: Instead of naming the muscles after mice (Latin: *mus* = mouse), we should name them after fish, for they do not behave like individual creatures darting here and there, but in flowing unity, not dissimilar to schools of fish in the water, streaming this way and that, separable but never

---

44 From a pedagogical point of view, we are aware that the student's limbs have been growing considerably in recent years, thereby increasing the challenges involved in coordinating the new range of their movements with the force necessary to create them, as well as the need to develop the muscle power required to provide the overall increases in force required by their increasingly lengthy Load Arms. It forces them to take hold of their bodies to a new degree, to "incarnate" more deeply.

Muscular System

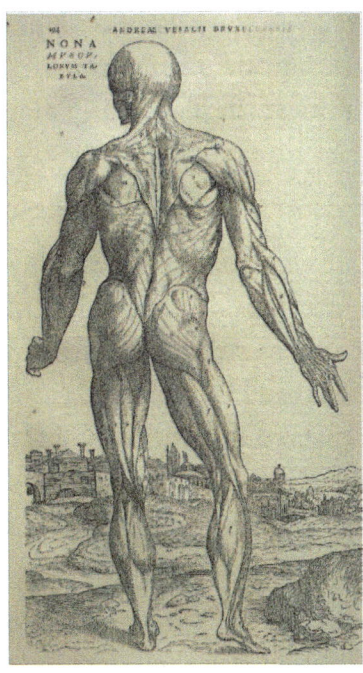

Fig. 1.57 Artistic rendering of human muscles, anterior view.
(By Andreas Vesalius, *De humani corporis fabrica* (1543) Wikimedia Commons)

alone! Muscles are always active in concert with others. In reference to the back muscles, for example, E.M. Kranich observes:

Even if differentiated into individual systems, the back muscles work as a whole. With every activity that a person undertakes with the back, with every bending or stretching of the trunk, by the smallest turn or sideways leaning, all the back muscles are involved. The individual muscle is a part of the overall whole that constitutes movement and posture. By itself, the individual muscle has no reality—just as an individual tone in music has no reality without its manifold relationships—the intervals—to the other tones. (2003, p. 30)

Others have compared the coordination of the different muscles during human movement with the performance of an orchestra (Bernstein 1988).

With this dynamic in mind, it is not surprising to learn that muscle tissue is roughly 79% water. This, together with its great affinity to the blood, has led some to describe the muscles as "formed blood" (Pressel 1984, p. 24). Of course this form is not solid, but in constant transformation. From the linear shape of a stretched muscle to the spherical tendency of a contracted one, activity calls more blood into the muscle and it swells up. With the relaxation that follows, much of it flows away. The athlete, the musician, the technician, all practice their essential movements so long until they achieve the harmony of movements they seek.

There are multiple ways to achieve this coordination of movements:

Smooth/involuntary muscles. The smooth muscles line most of the hollow passageways of the body. These muscles work constantly with wave-like (peristaltic) movements below the level of our consciousness. They play a vital role in a variety of the body's most basic functions, from tissue oxygenation to digestive processes. (These muscles will be covered in their full context in the 10th grade biology block.)

Cardiac muscle. The heart muscle—another 10th grade topic—has characteristics of both the smooth and the skeletal (striated) muscles. In the current context it is important to emphasize the never-ceasing rhythmical action of this muscle (hundreds of millions of rhythmical contractions during our lifetime). Here, too, the muscular activity is independent of our conscious control, even though we can influence it indirectly by doing things that increase our body's need for oxygen.

Through an introductory description of the two muscle types just mentioned, the students begin to realize that most fundamental biological processes (ones that keep us alive!) rely on the activity of muscle types that lie outside our conscious regulation (and largely our awareness). They begin to see that we simply presuppose their ongoing functioning for all that we do, day after day, night after night, for an entire lifetime!

Skeletal muscles of the lower limbs. As we saw earlier from the perspective of our bone structure, the lower limbs are focused primarily on support and locomotion, which require exceptionally strong extensor muscles (gluteus maximus, quadriceps, gastrocnemius/soleus) of the upper and lower legs to make upright walking possible. The foot muscles work more to secure the structure of the arch and stabilize the foot than to move the toes (Rohen 2007, Van De Graaff & Fox 1998).

Skeletal muscles of the upper limbs. The upper limbs, on the other hand, have been freed from the task of supporting the body and are capable of a great variety of movements. The muscles of the lower arm, for example, can turn it inward (pronation) to enable touching, holding and grasping, as well as outward (supination) to enable it to receive and carry objects. The pectoral girdle is so mobile that it gives the arms and hands a range of motion that can reach to almost any point in our field of vision. But the muscular activity of our upper limbs goes far beyond just engaging things in the material world. All that we find in outer culture, from simple tools to the highest forms of art, would not exist without the highly coordinated activity of our arms and hands. The movements of our arms and hands could not transform the thoughts and ideas we develop within into outer realities, if their movements were not permeated by a high level of consciousness. These high levels of thought-guided coordination are not inborn, of course, but result from countless hours of dedicated practice and the learning that accompanies it.

In such movements of the arms and hands, there are many muscles working together. To give the students a more concrete sense for this, it helps to provide them with some concrete numbers.

- Nine muscles are involved in the movement of the upper arm (humerus).
- Eight muscles move the forearm bones (radius and ulna).
- Fifteen muscles located in the forearm are engaged in the more powerful movements of wrist, hand, fingers and thumb.
- Ten intrinsic muscles that originate in the palm of the hand are involved in the more intricate and precise movements of the fingers.

Waves of contraction and expansion move through these muscles continuously as they work together to realize our intentions. The movements of the individual muscles are not significant in themselves, but only through their coordinated activity. They are subordinate to the overall movement that follows the intentions of the human being involved in a particular activity (Kranich 2003, Marieb & Hoehn 2012, Rohen 2007, Van De Graaff & Fox 1998).

The students are usually amazed to realize that they engage and school this orchestra of muscles, that they themselves are the conductors thereof! They have grown more and more adept at shaping the movements of their upper limbs and hands through childhood and—if they are so inclined—will continue to do so the rest of their lives! (After all, their teacher may still be trying to learn how to juggle, or to play the guitar, at age 55+.)

Gestures. There is yet another level, however, in which we normally engage our upper limbs without conscious intention: in our gestures. These "come naturally" to us and bring our inner life to expression. We can ask the students to demonstrate gestures that accompany different feelings such as joy, anger, rejection, welcoming, uncertainty, surprise. They are often fascinated to see how "non-subjective" they are, in the sense that we all share similar gestures in relationship to specific feelings. (Good acting is based on that fact.)

Muscles of facial expression: Our inner moods express themselves in yet another way as we move upward to the head region. Besides the muscles of the head that are involved in chewing (mastication)—which make use of leverage as we discussed earlier—we have a whole new type of organization in the muscles of facial expression. These muscles usually originate in the bones of the skull and insert into the skin or another muscle, but not into bone as the skeletal muscles do. They are thin, variable in strength and tend to fuse with adjacent muscles. In contrast to what we have discussed up until now, they move skin rather than joints! And in doing so, they express the fluctuations of our inner life with amazing dynamic and subtlety.

It is also fascinating to learn that specific inner experiences come to expression in the face through the contraction of very specific muscles.[45] Even though our inner experiences are private, personal, and often subjective in nature, their expression is quite objective at the bodily level.[46] That said, it should not surprise the students to learn that in addition to the primary muscles involved in a specific expression, a "background chorus" of muscles is also involved. For example, to create a smile that only lifts the upper lip and corners of the mouth, a minimum of 10 muscles must be engaged in differing degrees (Kranich 2003, Marieb & Hoehn 2012, Totoro & Derrickson 2013).

---

[45] Although it goes beyond the scope of a 9th grade block to go into great detail here, it is astounding to learn that our feeling life is connected to our muscle system in such a tangible way. Feelings such as amazement (epicranius), dissatisfaction (levator labii superioris), joy (zygomaticus major), sadness (depressor anguli oris), etc., are expressed very definitively through links to specific muscles (Kranich 2007).

[46] The students already understand through their acting experiences that, in order to be truly convincing in one's gestures and expressions, one must immerse oneself completely in the experiences of the character being depicted and feel strongly what they would be feeling. Without that immersion and corresponding feeling, the gestures and facial expressions become hollow.

When dealing with such matters, the teacher should be sensitive to the make-up of the class and how the students react to this topic. For the more extroverted ones, this can be a fun exploration into a plethora of facial expressions and the search for which muscles might be involved. For the class as a whole, however, it is not beneficial to have this topic lead them into excessive self-reflection about how "cute," "lame," etc., their own smiles, frowns, etc., are relative to their peers. At this age, in particular, such "self-absorption" should be avoided.[47] If it appears that discussing facial muscles might lead in this direction, I have found it best to touch on them briefly, and then to move on.

Larynx and Tongue. If we assume something like the sequence of topics as they have been presented here, the teacher can look back at this stage to an organ considered earlier—the larynx—and discuss how its muscles also serve the inner expression of the human being through voice creation and modulation.

It is fun to conclude our overview of the muscles with an amazing example: a muscle that works constantly for us, is always ready the moment we need it, and yet never seems to tire: the tongue. On the one hand it is its own muscle; it is made up of muscle fibers that squeeze, curl, and fold the tongue, thereby changing its shape with amazing speed and agility. These intrinsic tongue muscles are aided by extrinsic tongue muscles that anchor it and help it to protrude, retract, and elevate, as needed.[48] Its extreme flexibility and exactness of movement can be experienced by the students if we consciously produce various sounds of speech. Doing so, it immediately becomes obvious that the tongue works in unison with the muscles of the jaw and facial expression (as well as with the teeth, palate and

---

47 In the *Education for Adolescents*, Rudolf Steiner makes the important point that "if we look at the chief damages created by modern civilization, they arise primarily because people are far too concerned with themselves and do not usually spend the larger part of their leisure time in concern for the world, but busy themselves with how they feel and what gives them pain. ... And the least favorable time of life to be self-occupied in this way is during the ages between 14, 15 and 21 years old. ...The world must become so all-engrossing to young people that they simply do not turn their attention away from it long enough to be constantly occupied with themselves. For, as everyone knows, as far as subjective feelings are concerned, pain only becomes greater the more we think about it. It is not the objective damage but the pain of it that increases as we think more about it. In certain respects, the very best remedy for the overcoming of pain is to bring yourself, if you can, not to think about it" (*Education for Adolescents*, CW 302a, June 21, 1922).

48 The tongue's extreme flexibility is also enhanced by the fact that it has a movable base—the horseshoe-shaped hyoid bone—which is the only bone in the body that does not articulate directly with any other bone. Instead, it is simply anchored by ligaments to the styloid process of the temporal bones (Marieb & Hoehn 2012).

nasal cavity) in concert with the lungs and larynx to form speech. The letters "E" and "A," for example, are spoken by arching the tongue and slightly stretching the lips; "O" and "U" require lowering the jaw and pursing the lips; "Ah" is brought forth with an open mouth, relaxed lips and the tongue held low.

With some classes it can be fun to go through all 26 sounds of the alphabet and have everyone pick their favorite constellation of sound-forming-factors. Having the students slowly speak several sentences while paying close attention to the differentiated formative movements that take place in the organs of speech, can be an amazing experience for many of them. They also enjoy speaking the same sentences again at high speed to witness what an unbelievable capacity they possess.

I often conclude such exercises by pointing out that the human capacity for speech is a unique one. It is far different from the limited vocal expression available to animals, not only because it possesses a far greater variety of possibilities—which are not determined by heredity—but also because our speech can free itself from emotional constraints in order to serve as the instrument of language and conceptualization[49] (Kipp 2005).

With the constellation of muscles just discussed, we have reached a very rarified level in the expression of our inner life: We are not only able to communicate our thoughts and feelings in a differentiated way as they pertain to everyday life, but are capable of bringing to expression and communicating to others our highest ideals and our deepest intentions.

---

49 This perspective can be explored much more extensively in the 12th grade zoology block.

Bibliography – Part I

Ahrahams, M. (2012). Experiments show we quickly adjust to seeing everything upside-down. *The Guardian,* 12 Nov. 2012.

Barfield, O. (1988). *Saving the Appearances.* Middletown: Wesleyan Univ. Press, 26.

Benninghoff, A. and Goerttler, K. (1975). *Lehrbuch der Anatomie des Menschen, II. Band.* Munich: Urban & Schwarzenberg.

Bernstein, N. (1988). *Bewegungsphysiologie.* Leipzig: J.A. Barth (cited in Kranich 2007).

Biga, Dawson, et al. (2019). *Anatomy & Physiology.* Open Oregon State, Oregon State University.

Bohm, D. (1981). *Wholeness and the Implicit Order.* London, Boston: Routledge & Kegan Paul.

Crick, F. (1994). *The Astonishing Hypothesis. The Scientist's Search for the Soul.* New York: Charles Scribner's Sons.

Frankl, V. (1969). Reductionism and Nihilism. In Koestler, A. and Smythies, J. *Beyond Reductionism.* Boston: Beacon Press, 398.

Gao, L.; Balakrishnan, S.; He, E.; Yan, Z. and Müller, R. (2011). Ear Deformations Give Bats a Physical Mechanism for Fast Adaptation of Ultrasonic Beam Patterns. *Physical Review Letters,* 2011; 107 (21).

Glas, N. (1976). *Gefährdung und Heilung der Sinne.* Stuttgart: J.C. Mellinger Verlag.

_____. (2004). The Feet Reveal the Will. *Waldorf Journal Project* #3. AWSNA Publications.

Goethe, J.W. (1970). *Theory of Colours.* Cambridge: MIT Press.

Hanson, N. (1958). *Patterns of Discovery.* Cambridge: Cambridge Univ. Press.

Heusser, P. (2016). *Anthroposophy and Science.* Frankfurt: Peter Lang, Edition.

Julius, F. (1984). *Entwurf einer Optik.* Stuttgart: Freies Geistesleben.

Kahnemann, D. (1973). *Attention and Effort.* Englewood Cliffs: Prentice-Hall.

Kegan, R. (1982). *The Evolving Self.* Cambridge: Harvard Univ. Press.

Kiefer, B. (2007). Das Prinzip der Emergence. Schweizerische Nationalfonds. *Horizonte,* 33. Cited in Heusser, P. (2016). *Anthroposophy and Science.* Frankfurt: Peter Lang, Edition, 83.

Kipp, F. (2005). *Childhood and Human Evolution.* Hillsdale, NY: Adonis Press.

Koenig, K. (2006). *A Living Physiology.* Camphill Books.

Kranich, E. (2003). *Der Innere Mensch und Sein Leib.* Stuttgart: Freies Geistesleben.

Leber, S. (1993). *Der Menschenkunde der Waldorfpädagogik.* Stuttgart: Verlag Freies Geistesleben.

Marieb, E. and Hoehn, K. (2012). *Human Anatomy and Physiology.* San Francisco: Pearson.

Mörike, K.; Betz, E. and Mergenthaler, W. (2001). *Biologie des Menschen.* Heidelberg: Quelle u. Meyer.

Piaget, J. and Inhelder, B. (2000). *The Psychology of the Child.* NY: Basic Books.

Pressel, S. (1984). *Bewegung is Heilung.* Stuttgart: Freies Geistesleben.

Raven et al. (2008). *Biology.* New York: McGraw-Hill.

Richter, T. (2016). *Pädagogischer Auftrag und Unterrichtsziele.* Stuttgart: Verlag Freies Geistesleben.

Rohen, J. (1978). *Funktionelle Anatomie des Nervensystems.* Stuttgart: Schattauer Verlag.

_____. (2007). *Functional Morphology.* Hillsdale, NY: Adonis Press.

Rosslenbroich, B. (2019). Eigenschaften des Lebendigen, in Zimmerman & Wallmann: *Biologie in der Waldorfschule.* Stuttgart: Verlag Freies Geistesleben.

Rucci, M. and Beck, J. (2005). *Effects of ISI and flash duration on the identification of briefly flashed stimuli,* Spatial Vision. 18(2), 259–274.

Schumacher, E. (1977). *A Guide for the Perplexed.* NY: Harper & Row, 17.

Science Buddies. (2015). Ears: Do Their Design, Size and Shape Matter? *Scientific American,* Nov. 2015. https://www.scientificamerican.com

Sloan, D. (2018). "Beyond the Mechanistic World View." *Research Bulletin, Vol. 23/1*. Hudson, NY: Waldorf Publications, 5–16.

Steiner, R. (1988). *The Science of Knowing*. Mercury Press. Ch. 11.

_____. (1996). *Education for Adolescents* (CW 302). Great Barrington, MA: Anthroposophic Press.

_____. (GA 305) *The Spiritual Ground of Education*. Lecture VI. Aug. 22, 1922. Hudson, NY: SteinerBooks.

Strauss, E. (1966). *Phenomenological Psychology* NY: Basic Books.

Tatler, B.; Wade, N.; Kwan, H.; Findlay, J. and Velichkovsky, B. (2010). Yarbus, eye movements, and vision. *Iperception*, 2010; 1(1): 7–27.

Totoro, G. and Derrickson, B. (2013). *Principles of Anatomy and Physiology*. Hoboken, NJ: John Wiley & Sons.

Van De Graaff, K. and Fox, F. (1998). *Concepts of Human Anatomy and Physiology*. Chicago: W.C. Brown.

Von Mackensen, M.; Allgöwer, S. and Bielfeld-Achermann, A. (2004). *Uprightness, Weight and Balance*. Fair Oaks, CA: AWSNA Publications.

Wagenschein, M. (1982). *Teaching to Understand—On the Concept of the Exemplary in Teaching*. https://natureinstitute.org/txt/mw/exemplary.pdf. (For several additional essays by Wagenschein on exemplary teaching, go to the Nature Institute Website.)

Whitehead, A. (1967). *Science and the Modern World*. New York: Free Press.

Yarbus, A.L. (1967). *Eye Movements and Vision*. New York: Plenum Press, 1967.

INTERMEZZO

Waldorf Biology in a Reductionist Setting

Intermezzo

Most biology teaching in the United States today is based on a reductive-analytic approach to the life sciences. As one well-respected biology textbook says in its introductory chapter (Raven et al. 2016):

> *Scientists often use the philosophical approach of* **reductionism** *to understand a complex system by reducing it to its working parts. Reductionism has been the general approach of biochemistry, which has been enormously successful at unraveling the complexity of cellular metabolism by concentrating on individual pathways and specific enzymes. By analyzing all of the pathways and their components, scientists now have an overall picture of the metabolism of cells.*

After breaking down the organism into its smallest entities, the many functions of biological systems are seen as

> *…both determined and constrained by the principles of chemistry and physics. … Every level of biological organization is governed by the nature of energy transactions learned from the study of thermodynamics.* (p. 7)

Raven et al. conclude their introductory chapter on "The Science of Biology" with the following statement:

> *Biology as a science is broad and complex, but some unifying themes help to organize this complexity. Cells are the basic unit of life, and they are information-processing machines.* (p. 14)

This introduction is followed by a chapter on "The Nature of Molecules" and one on "The Chemical Building Blocks of Life." The textbook then devotes 335 pages to the biology of the cell and its genetic and molecular manifestations. All of this before the themes of evolution, plants and animals and, lastly, ecology are considered. Such an organizational sequence and emphasis in a modern textbook on biology will surprise no one who has taken such a course in the last 50 years. This is the way that students in high schools and universities across the country are introduced to this field of study. The reductionist approach maintains that the parts are not only prior to the whole, but are, lastly, more real. Any form of integrative, holistic thinking is given, at best, secondary status. This fragmentary view extends far beyond biology textbooks. It is deeply rooted in the habits and attitudes of the modern science in general. As the distinguished physicist David Bohm put it:

> *Of course, the prevailing tendency in science ... tends very strongly to re-enforce the general fragmenting approach because it gives men a picture of the whole world as constituted of nothing but an aggregate of separately existent atomic building blocks, and provides experimental evidence from which is drawn the conclusion that this view is necessary and inevitable. In this way, people are led to feel that fragmentation is nothing but an expression of "the way everything really is" and that anything else is impossible. So there is very little disposition to look for evidence to the contrary. ... Even when such evidence does arise... the general tendency is to minimize its significance or even to ignore it altogether.* (1981, p. 15)

Although such a reductionist view of the world has proven very effective at certain levels of existence, if taken seriously as the primary basis for viewing all existence, it leads to a picture of nature that is—in the words of the eminent philosopher and mathematician Alfred North Whitehead—"a dull affair, soundless, scentless, colorless, merely the hurrying of material, endlessly, meaninglessly" (1969, p. 54).

Whitehead also points to another kind of conclusion that can follow from the reductionist understanding of the laws of nature.[1]

> *The human body is a collection of molecules. Therefore, the human body blindly runs, and therefore there can be no individual responsibility for the actions of the body. If you once accept that the molecule is definitely determined to be what it is, independently of any determination by reason of the total organism of the body, and if you further admit that the blind run is settled by the general mechanical laws, there can be no escape from this conclusion.* (p. 78)

If we continue in this vein and take one of the fathers of modern genetics, Francis Crick, seriously, then when we are speaking of the human being we

---

[1] Whitehead is not saying that the above description represents his own view of the relationship between mind and body.

Intermezzo

assume a complex system that is explained by the behavior of its parts (which, in turn, have to be explained by the properties of the "parts of those parts" and how they interact). To understand the brain, for example, we need to know not only the interaction of the neurons themselves, but these also need to be explained in terms of the ions and molecules of which they are composed. If we see this as a sufficient means for understanding the brain then we must agree with Crick that we are *"nothing but a pack of neurons."* In other words, our joys and sorrows, our sense of personal identity, our memories and ambitions are "in fact, no more than the behavior of a vast assembly of nerve cells and their associated molecules"[2] (1994, p. 3).

Seen from a pedagogical perspective, such a view—if taken seriously by students at the high school or university level—can lead to the conclusion that life is, in fact, meaningless. The obvious question for them becomes "What's the point in this life we live, if reality is only matter in motion?"

And even if they do not think things through to this ultimate conclusion, their picture of nature remains one of a lifeless interaction at the level of molecules. This has consequences for the way we relate to the world around us. In his thought-provoking article, "Beyond the Mechanistic World View," Douglas Sloan describes how the mechanistic picture of the natural world,

> *by removing the holistic view of a meaningful and valuable picture of nature has led to a relentless dismantling of nature… the view of nature as nothing but matter in motion also supports the exploitation and misuse of the earth through an unrestrained economism…The costs to the earth are now painfully apparent. The destruction of forests; the degrading of arable land; the pollution of lakes; the mass extinction of living species. … As long as nature is regarded as having no qualities— no inner life, no meaning, no living wholeness—taking it apart for our own immediate pleasure and economic advantage is obviously that much easier to justify.* (2018, p. 14)

---

[2] That such a description undermines itself has been pointed out by numerous more reflective, philosophically-schooled individuals. In the words of Douglas Sloan (2018): "The paradox in such writing lies in the fact that the ideas, values, and positions advanced by these scientists and thinkers must also be regarded as 'electrochemical brain processes,' thereby losing any qualitative advantage over other ideas, values, and position, all reduced to the same level of electrochemical mechanism. Either these thinkers are making exceptions for their own ideas or they are unaware of the implications of the mechanistic view so deeply ingrained in the modern scientific mind" (p. 10).

Sloan goes on to describe the tragic consequences that this mindset leads to in the widely prevalent factory farming of animals, where the suffering of animals—above all cows, pigs and poultry—is almost never addressed.

> *Daily our culture inflicts cruelty and suffering on millions of animals of an intensity hitherto unknown. The animals are defined as "units of production" and are treated accordingly as useful pieces of machinery without feelings. Their entire lives are unrelieved wretchedness. A pall of suffering of living, feeling creatures hangs over our modern culture, and most of us are complicit in it, if only through willful ignorance of what is taking place. ... The withholding of mercy to these fellow creatures bespeaks an appalling failure of imagination in thinking, a lack of empathy in feeling, and a weakness in moral willing.*

Of course, the reductionist, mechanistic approach is only one possible way to investigate nature. The poignant question is whether this way of viewing nature suffices for grasping all levels of existence—or does it, already from the outset, exclude certain aspects of nature that are actually characteristic of living organisms. Might it be that this approach is not completely false, but just extremely one-sided and thus unable to think outside its own self-created box, which makes it incapable of recognizing different levels of existence, such as plant, animal, and human?

The eminent psychiatrist, Viktor Frankl, characterized this reduction of phenomena to what only fits into the narrow lens of the specialist as *"nothing-but-ness."* Thus, for Francis Crick humans are—as cited earlier—"nothing but" a pack of neurons. Through limited perspectives such as this, as Frankl put it: "Human phenomena are thus turned into mere epiphenomena" (1969, p. 398).

E.F. Schumacher provides an original and insightful analysis of such issues in his book, *A Guide to the Perplexed* (1977, p. 17), where he speaks of the inability of reductionist approaches to distinguish between what he calls "ontological discontinuities" or "jumps in the Level of Being"—such as the fundamental difference between a stone, a plant and an animal. From a rigid reductionist perspective—he humorously observes—a dog could be defined as a "barking plant" or as a "running cabbage." As farfetched as this may sound, it is common in the realm of human biology, for example, to simply refer to the heart as "nothing but" a pressure-propulsion pump.[3]

Intermezzo

One attempt to deal with the challenge of "ontological discontinuities" can be found in the concept of "emergence," which points to the limitations of exclusively reductionist explanations when it comes to higher levels of complexity in nature. In the periodical, *Horizons*, published by the Swiss National Science Foundation, we find the following description of "emergence":

> *Emergence is one of the most puzzling but fundamental phenomena of the universe: the appearance of new characteristics at each higher level of complexity which cannot be predicted from the previous level. An example: The characteristics of life cannot be deduced from lifeless matter. However far we pursue research in physics and chemistry, this route will never enable us to predict the specific behavior of living organisms. It appears to be a generally valid principle that the (complex) whole cannot be traced back to its (simple) parts. This includes all stages of increasing complexity. At the level of the atom: Observing hydrogen and oxygen atoms in isolation gives no clue to the characteristics of a water molecule. Or, at the end of the scale: The characteristics of consciousness do not result from the extrapolation of behavior.* (Kiefer 2007)

Teaching the life sciences in a Waldorf high school includes an awareness that questions around reductionism, "nothing-but-ness," ontological discontinuities, etc., are not trivial. Fortunately, a great deal of significant research from a holistic/phenomenological perspective has been done in the life sciences over the past century. These efforts provide Waldorf science teachers with many new and exciting vantage points to draw upon as they try to nourish the "emerging capacities" of their students moving through the four years of high school. Many of these resources will be referred to in what follows, because they provide a means for fostering new kinds of thinking in our students, thinking that allows them to move beyond mere "nothing-but-ness" and linear causality into a more dynamic and multifaceted understanding of the nature of nature.

---

3   This is a central topic that we will discuss in detail in the context of the 10th grade biology block.

Bibliography: INTERMEZZO

Bohm, D. (1981). *Wholeness and the Implicate Order*. London, Boston: Routledge & Kegan Paul.

Crick, F. (1994). *The Astonishing Hypothesis: The Scientific Search for the Soul*. New York: Charles Scribner's Sons.

Frankl, V. (1969). Reductionism and Nihilism. In Koestler, A & Smythies, J. *Beyond Reductionism*. Boston: Beacon Press, 398.

Kiefer, B. (2007). Das Prinzip der Emergence. Schweizerische Nationalfonds. *Horizonte,* 33. Cited in Heusser, P. (2016). *Anthroposophy and Science*. Frankfurt: Peter Lang, Edition, 83.

Raven et al. (2008). *Biology*. New York: McGraw-Hill.

Rosslenbroich, B. (2019). Eigenschaften des Lebendigen, in Zimmerman & Wallmann: *Biologie in der Waldorfschule*. Stuttgart: Verlag Freies Geistesleben.

Schumacher, E. (1977). *A Guide for the Perplexed*. New York: Harper & Row. 17.

Sloan, D. (2018). "Beyond the Mechanistic World View." *Research Bulletin, Vol. 23/1.* Hudson, NY: Waldorf Publications, 5–16.

Whitehead, A. (1967). *Science and the Modern World*. New York: Free Press.

Part II

Tenth Grade Human Biology

Part II. 10th Grade

Process-based Understanding

In the 10th grade biology block we move beyond the straightforward causal-connections emphasized in the 9th grade. Our task is to exercise and school a dynamic kind of thinking that can provide the basis for the formation of more complex and multifaceted judgments. In this way we help the students move beyond the "nothing-but" judgments characterized by Viktor Frankl to more comprehensive perspectives. After all, most of our students—whether they attended the Waldorf grade school or not—enter this block with the view that the heart is "nothing but" a pump, and the brain "nothing but" a high-level computer.

The streaming, flowing dynamic found in our internal organs provides excellent material for overcoming such overly simplistic and mechanistic views of the human being. Although we normally do not bring it to awareness, a "process-based understanding" of the human organism is closer to reality than the static, "finished" picture we normally carry of it. As Rudolf Steiner has pointed out:

> *The outer spatial forms of human organs are merely living processes that have come to a "standstill" for a moment. In reality, organs such as the lungs, stomach, heart, liver, & kidneys are not that which they appear to be at first glance: as clearly circumscribed, quiescent forms. No, these organs only give the illusion of such solidly and constancy, for in reality they are in constant living movement, they are not finished, completed forms, but living processes. We should speak of the heart process, the lung processes, kidney processes, etc.* (CW 79, p. 68)

Only for our outer perception do the organs of the body appear so finished. In reality, even our solid bones form out of our fluid organization and are constantly restructuring based on the kinds of stress put on them. We do not, of course, present this fact to the students so directly, but try to gradually awaken in them a sense for the process-nature of all living things in contrast to the snapshot, finished impressions that we normally have of them.[1] As indicated earlier,

---

1 In a very general sense, one can say that the latter perspective is emphasized in Anatomy, while the former predominates in Physiology.

this involves developing a sense for the complexity with which the processes within living organisms are interwoven and requires a mobility of thinking that goes beyond merely grasping the kind of linear causality we intentionally emphasized in the 9th grade block.

Building upon such a dynamic understanding of the organism, it is then possible to relate what we find on the bodily level to the inner life of the human being. A second goal of this block is to shed light on the multifaceted connections that exist between that aspect of our being that is sense perceptible to all (the body), and that which only we ourselves can perceive (our inner experiences).[2] The inner life of 10th grade students is usually a very active and frequently a very subject-oriented one. Bringing more objectivity to some aspects of those inner experiences is often helpful to them at this phase of their lives. In the 9th grade we touched briefly on the inner life through references to gestures and facial expressions. But now, in the 10th grade, the relationship between the inner and the outer human being can be explored at a more subtle and significant level that helps transcend the common view that our consciousness is "nothing but" nerve/brain activity.

---

2   As described by Rudolf Steiner in his discussions with the faculty of the Stuttgart Waldorf School, the task of the 10th grade biology block is to "help the students understand the individual human being ... through the study of the organs and organ functions of the physical body in connection with the soul and spiritual" (CW 300b, p. 270).

Part II. 10th Grade

Getting Started

Similar to the 9th grade biology block, I find it worthwhile to start broadly, thereby creating a context for what will follow. One way to do this is to briefly refer to the 9th grade and the topics covered. I remind the class that we dealt with the so-called "outer organs" in that block and will now move on to the "inner organs," ones that are not directly perceptible to us and do not allow (for the most part) our direct influence. From there I ask the students to make a list of all the organs that they have heard of, which can then be gathered and written on the board. I then ask the class: If we imagine all these organs (and any that might be missing in our list) working together in a healthy way, have we captured the whole human being?

After a few moments of reflection, someone will say: "But what about our memories, what about our feelings?" Of course others will agree very quickly: Yes, our inner life is a huge part of what it means to be a human being and must be considered too. From there the teacher can point out how special fields of inquiry have been developed that focus on various aspects of the inner and the outer human being. From the bodily perspective, we have anatomy and physiology, but also the study of heredity, biochemistry, etc. The inner human being is studied from various angles in the fields of psychology, sociology, philosophy, etc. The connections between the inner and outer are also considered to varying degrees, for example in psychiatry, which links the medical and the psychological.

In laying the groundwork for this block, it is important to be clear that when we talk about our inner life, we are speaking about a realm that—in clear contrast to the bodily level—cannot be perceived with the senses: It cannot be seen, touched, heard, smelled, tasted… A few simple examples can follow. We can ask one student to picture what they had for breakfast. The rest of the class observes them carefully as they are doing this. Do we now know whether this student had toast with their scrambled eggs, or not? Or we ask the class what color the teacher is focusing on at this moment as they look at a painting on the back wall of the classroom. In this context, I often also tell the students how amazed I was as a seventh grader riding to school one day in a car full of kids, when it suddenly dawned on me that neither they nor the parent who was driving could perceive (and hence know) what I was thinking! My inner life was accessible to me and only me! What a liberating—but also lonely—moment of discovery!

In this and other ways we can try to help them see that we are dealing with two very different aspects of experience here, and that the relationship between the two—frequently known as "the mind-body problem"—has interested philosophers for centuries. Psychosomatic medicine (*psycho* = mind; *somatic* = body) is another example of a field of study that focuses on this relationship.

It may dawn on the students, in the course of such a discussion, that last year we spoke a great deal about our inner experiences when studying the sense organs. We not only spoke of the eyes, but also of seeing; we studied the ear, but always in the context of hearing, etc. Seeing and hearing are part of our inner life, and as such are not perceptible to others. No one else can see our seeing or hear our hearing: These experiences are not body; they are not part of the outer, sense-perceptible world. Even the world's most skilled and experienced brain surgeons have never come across a thought or a feeling when investigating the depths of that organ.

From this perspective, our bodily sense organs can be viewed as instruments of our inner life, as instruments by means of which our inner life has access to and can experience the outer world. It is, therefore, no surprise that the word "organ" comes from the from the Ancient Greek word ὄργανον (órganon), which means an instrument, an implement, a tool.[3] Of course, we also possess bodily instruments through which our inner being can express itself and become active in the outer world. This is what our limbs do for us, our skeletal/muscular system that we also studied last year. It serves us at many levels, from moving through space all the way to the expression of our inner life through the organs of speech.

In the course of this current block, we tell the students, we will want to be alert to other, less obvious "organs" that also serve as instruments of our inner life in one form or another.

---

3   Whereas we normally bring a new concept that sheds light on the phenomena we are studying the day after the new material has been introduced, in this case we did not let them "sleep on it" for just one night, but for a whole year!

Part II. 10th Grade

The Cardiovascular System (CVS)

A central objective of this block is to help students develop a more fluid, dynamic thinking that is not satisfied with the easy, quick, one-dimensional judgments that are so common in life today. Most things have many aspects to them and an important part of our task is to help the high school students move beyond the "multiple choice, one right answer" view of the world. We hope to engender the expectation that the things we encounter in life can be looked at from various perspectives—and must be—if we are to get below the surface and really understand them. For such a venture, the cardiovascular system (CVS) provides an excellent starting point. The flow of the blood throughout the body immediately involves us in essential processes that are in constant transformation. To understand the flow of the blood, for example, the students will need to develop dynamic mental pictures and recognize multiple interpenetrating contributory factors, which will necessitate—as we shall see—that they move beyond a simple "nothing but" causal picture—such as the common reference to the heart as "*only* a pump."[4]

In what follows, more material will be presented than can be covered in a normal biology block. I have done this intentionally because, depending on the class in front of one, the timing and length of the block, and other variables, it is very helpful when teaching a certain subject area to have a number of possible approaches and aspects of the topic to choose from. In this way the teacher can customize the lessons to more effectively meet the here-and-now as the block progresses.

One possible way to begin this investigation is to pose the age-old question: What came first, the chicken or the egg, but with a slight variation: What came first, the flowing blood or the beating heart? For most students this is not even

---

[4] That the "heart as pump" model needs revision not only in educational contexts, but also in the practice of mainstream medicine, is evidenced by the publication of Prof. Dr. Branko Furst's pioneering book, *The Heart and Circulation—An Integrative Model* (2014, 2020). With several introductions that emphasize the significance of the book by professors of medicine from Harvard University, the Mayo Clinic, and the German Heart Center in Berlin, Professor Furst has produced a keystone work on the nature of the human cardiovascular system, one which shows the limitations of the "heart as pump" model and provides a much broader view of the central role played by the peripheral circulatory system.

a real question: Of course the heart comes first—it causes the blood to flow! Without going into great detail, it is worthwhile to begin with a few anecdotal examples that make the question more concrete. For example, it is interesting to note that in the periphery of the human embryo, blood islands form and then differentiate into two different cell types: the flattened vessel cells that move to the outside and unite to form the tubular blood vessel and the future blood cells that remain on the inside and flow down the tubes created by the vessel cells.

This happens at a time when the heart is still forming. It could be explored in more detail, of course, but rather than entering into the complexity of the human embryonic development,[5] a few additional findings can be mentioned that provide a first taste of the surprising nature of this topic without bringing it to a conclusion (to a "nothing but" answer).

Researchers have found, for example, that frog embryos live for up to two weeks after the primordial heart has been removed. Other experiments have shown that Mexican salamanders with defective hearts that don't beat, show normal swimming behavior and survive for up to two weeks, at which time the nutrients from their yolk sac run out. One researcher removed the tube heart of a bullfrog larva, turned it 180 degrees and re-implanted it so that its contractions went in the opposite direction of (= against) the blood flow. To his amazement, the blood continued to flow in the direction it always had—from the arterial to the venous vessels (Furst 2020, pp. 37–38)!

A more extreme example—and perhaps a disagreeable one to modern sensibilities—is a series of experiments done by Manteuffel-Szoege[6] and colleagues in the 1960s involving asphyxiated dogs. Once an asphyxiated dog's heart had stopped for 10–15 minutes, the researchers allowed a flow of oxygen to enter the lungs through a tube. Although the blood flowed much less rapidly and without measurable pressure, it continued to circulate for two hours after the heart had been stopped—at which point the researchers ended the experiments, although fully expecting that the blood flow would continue on for much longer still (Furst 2020, Manteuffel-Szoege 1977).

---

5   Many Waldorf schools have an independent main lesson block in the 10th (or 11th) grade that deals with human embryology in detail. For more to this topic, see Holdrege (2014).
6   Manteuffel-Szoege was the lead physician for surgery at the University Clinic of Warsaw, Poland, and played a central role in introducing modern methods of heart and lung surgery in Poland.

Although such anecdotal examples are extreme and violate modern sensibilities, since they have taken place in the past we can still learn from them. They can serve to "get a foot in the door" with students for whom, until now, "the heart is a pump" has lived as an unquestioned assumption that needs no further consideration, while at the same time providing the basis for a discussion about the serious ethical issues around animal experimentation as they have come to be viewed in our time.

Circulation—the Flow

At this point, whether we have opened some minds or not, we need to get the blood flowing. In the spirit of starting with "the whole" and waking up gradually—through a process of analysis—to the parts, it is helpful to take a quick look at a circulatory system that really is "circular." The fish can provide this for us.

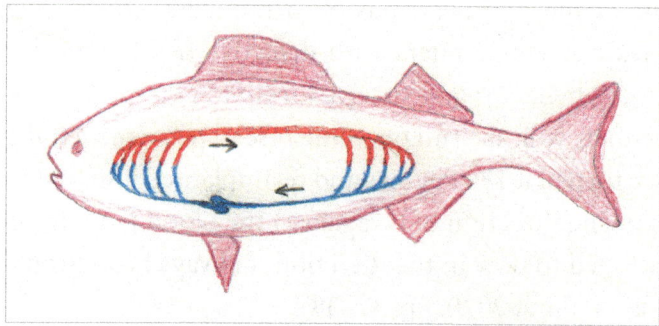

Fig. 2.1 Simplified diagram of fish circulation.

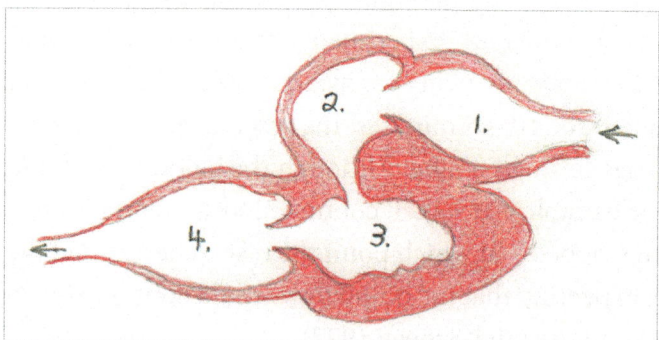

Fig. 2.2 The venous heart of a fish. 1. Sinus venosus. 2. Atrium. 3. Ventricle. 4. Bulbus arteriousus.

With the help of easy-to-understand drawings, we get the students' minds into movement as they picture sequentially the stages of the blood's flow, starting with the blood's passage through the simple S-shaped cardiac tube that makes up the fish's heart,[7] to the capillaries in the gills (where the blood is oxygenated), then out into the dorsal aorta, from which the blood is distributed through the capillaries in the body organs, before descending to the heart and then rising again in the gills. The goal is to get the student's perception-guided imagination into movement rather than falling into the typical stop/start mode of isolating, identifying and naming anatomical parts, while losing a sense for the dynamic of the whole (Hickman et al. 2001, Rohen 2007).

If we simplify this "circular flow" to a diagram, we get something like this.

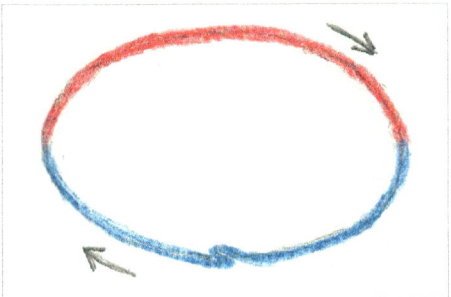

Fig. 2.3 Simplified circular blood flow pattern with the heart in the periphery.

The next question becomes, how does human blood flow fit into this picture. After some discussion we conclude that the circle does not suffice, that we have a "double circulation"—a systemic and pulmonary circuit with the heart in the center. We then attempt a simple diagram on the board that creates a unique loop for each lung as well as one for the head and one for the rest of the body.

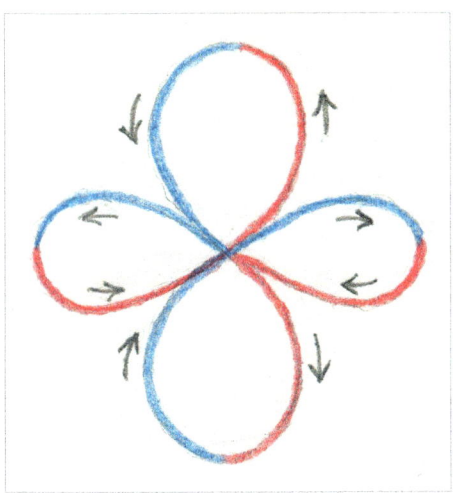

Fig. 2.4 Simplified pattern of the blood flow with the heart in the center between the systematic and pulmonary circuits.

---

7   A fish heart consists of a sequence of four compartments through which pass constant waves of peristaltic contractions.

Part II. 10th Grade

The task now is to bring this simple form into movement (flow). Once the students can "flow" through this diagram with ease, while formulating in a general manner what is happening with the blood as we move through it, we can provide a more differentiated drawing with some basic labels, where we also take into account that the heart is not one muscular tube as in the fish, but divided into two halves. The reason for this becomes obvious when we follow the flow. We practice the sequence, the students move through it using arm movements. If there is enough space in the classroom,[8] we trace it on the floor and walk through two at a time, with their classmates observing carefully to see

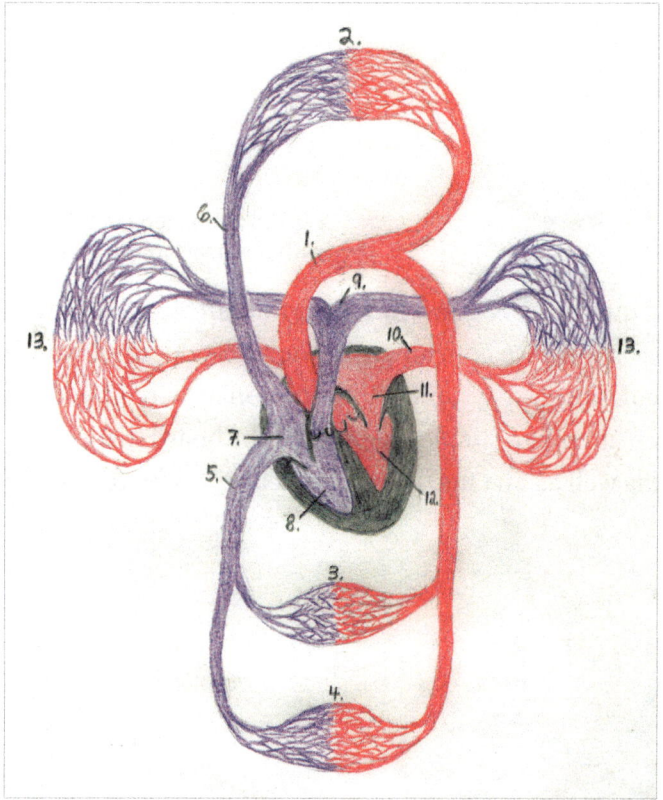

Fig. 2.5 Diagram of blood flow in the systemic and pulmonary circuits with a four-chambered heart. 1. Aorta. 2. Capillaries in the head and neck. 3. Capillaries in the internal organs. 4. Capillaries in the legs. 5. Inferior vena cava. 6. Superior vena cava. 7. Right atrium. 8. Right ventricle. 9. Pulmonary artery. 10. Pulmonary vein. 11. Left atrium. 12. Left ventricle. 13. Lung capillaries.

---

8 If there is not enough space, we go to the gym or another common space.

if they get it right! This can be a somewhat chaotic, but a fun moment and helps them internalize the dynamic. After returning to their seats, they add arrows to their drawings showing the direction of the flow throughout the diagram. Part of their homework will be to describe the path of the blood flow from when it enters the foot, until it returns to the foot again (for the sake of this exercise, we presume that it will return there and not elsewhere).[9]

Although this may seem to be a disproportionate emphasis on something that appears quite straightforward, I have been amazed over the years to see how difficult many students find it to accurately recreate this flow sequence in their imagination—even with the help of a drawing and with the levels of oxygenation indicated by red and blue lines. In an age of screens, when the students are accustomed to moving from finished image to finished image without any effort of their own, or where they simply follow the flow of moving pictures (video, etc.), the use of their own will to create an "accurate sensorial flow-imagination" calls on a capacity they will need to practice and strengthen if they are to do justice to living nature.

Arteries and Veins

Although it would be easy to move from what has just been discussed straight to the heart, I have found it worthwhile to stay with the flow a bit longer and to emphasize next the qualitative differences between the fluid dynamic in the arteries and veins. Our drawings up to now do not show this. The only difference that is symbolized by the red and blue lines is the ratio of CO_2 to O_2 and the nutrient and metabolic-waste content of the blood in those vessels. To gain a more qualitative sense for the differences, a simple comparison between a mountain stream and a river near the end of its journey can help. The mountain stream is daunting in its powerful dynamic: Bubbling and frothing with oxygen-permeated white water, it rushes down the steep mountainside, turbulent and ever-changing as it erodes away the earth beneath it.[10] A few anecdotes help

---

[9] It is usually necessary to remind many of the students at this juncture, that "right" and "left" in anatomical illustrations are always determined from the perspective of the object (drawing) in question. This means that, from the viewpoint of the person looking at the chalkboard, what is on the left side of their visual field is considered to be the right side of the drawing and vice versa.

[10] The high O_2 content in our arterial blood is also that which enables the substance transforming and consuming processes in bodily tissues.

here. Often the teacher or a few students can add their own of stories of trying to cross a raging, unbridled current in the spring, or can describe the kind of expertise it takes to kayak down such a turbulent flow.

Very different are the conditions far downstream, where many tributaries have gathered to form the slow-moving, dark and sediment-rich river that is nearing the end of its journey, soon to empty its load into a body of still water, often forming new land—even a delta—in the process. Some of the students will have swum in such a river and can describe what it is like. Personal stories such as my experiences swimming in the River Seine, where certain unseemly metabolic end-products from living creatures would sometimes float by (we learned not to swim with our mouths open!), might be considered "gross" by some, but the feelings that are triggered by such tales imprint the topic much more strongly into the students' memories.[11]

Returning to human circulation, the powerful flow of the arterial blood as it leaves the aorta is oxygen-rich ("white water"), and permeated by pulsing pressure waves as it rushes out into the periphery. Everyone knows, if you cut a large artery, the blood spurts out in bursts so powerful that they can only be stopped by cutting them off upstream with the help of a tourniquet. The flow in the veins, on the other hand, is much more like the Mississippi: mellow, voluminous, low in oxygen, but rich in CO_2 and waste material. No more pulsing or turbulence here. And like the downstream river, it is in the veins that we find most of the body's blood. Passing rapidly downstream, the arteries contain only 15% of the body's blood, whereas the other 85% gathers in the mellow flow of the so-called low-pressure system.[12]

With this in mind, it is not surprising to learn that the veins are thin-walled with large lumens that can accommodate greater volumes of blood than the strong-walled, elastic and muscular arteries that change their shape very little in response to variances in blood flow (Marieb & Hoehn 2012).[13]

---

11 As Rudolf Steiner points out in *Education for Adolescents* (CW 302), lecture one.
12 The low-pressure system includes the venous system, the capillaries, and the pulmonary system (Brettschnieder 2002).
13 The large arteries near the heart are the most elastic; the more distal distributing arteries are more muscular.

The Cardiovascular System

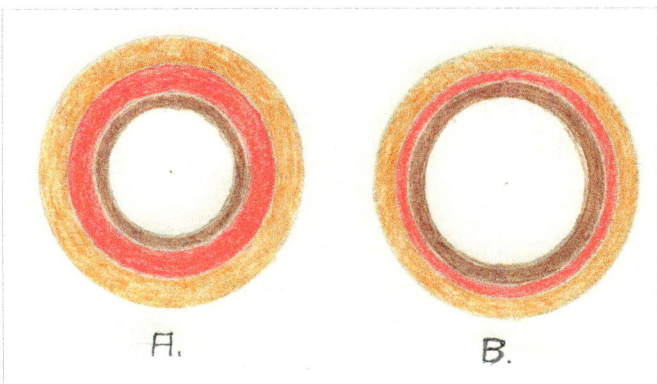

Fig. 2.6 Vertical sections comparing arteries and veins.
A. Arteries have a much thicker middle layer of smooth muscle (tunica media) and a smaller lumen. **B.** Veins have less smooth muscle and a larger lumen.

The veins also have a little something extra. When they pass between skeletal muscle groups, the contraction of the muscles creates a massaging effect on the them that helps move the blood along, but only in one direction: toward the heart. This is thanks to one-way valves found in the veins located in this region, as the drawing below illustrates.

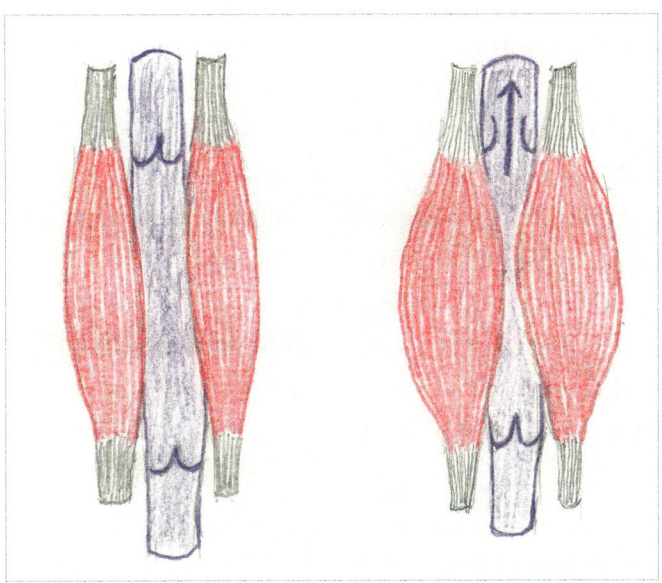

Fig. 2.7 The contraction of skeletal muscles helps to move blood toward the heart. The blood is prevented from moving away from the heart by the closure of the venous valves.

Part II. 10th Grade

To experience the blood flow in the veins, the students can place a finger on a distended blood vessel of the back of the hand and move it along, flattening the vein until the finger is released. In this way they can explore the network of blood vessels accessible there.

Capillaries

Between the arteries and veins we enter a new world, literally a new dimension. As the arteries branch out more and more on their journey to the periphery, they grow smaller and smaller before disappearing at a certain point![14] For the unaided eye the circulating blood seems to vanish and become one with the tissues and organs it enters, only to appear again for the observer at the other end—so to speak—when the tiny veins (venules) combine and become larger and larger veins on the way to the heart. With the help of a microscope we can determine that in this invisible region—that contains interweaving networks of very, very, very, tiny blood vessels known as the capillaries—the blood gives up its vessel-bound isolation and gives itself over to the needs of the organs and tissue around it. The blood plasma and white blood cells[15] flow out through the clefts and pores in the walls of the capillaries and permeate the extracellular spaces around them in a process known as microcirculation. Nutrients and oxygen are brought to the organs, carbon dioxide and wastes are carried away. The blood flow also cools organs that are producing surplus heat and warms those with less than optimal warmth. Although capillaries are found near to almost every cell in the body, their density depends on the metabolic activity of the tissues they serve. At any given moment the blood flows into only about 25% of the capillaries. The relative amount of blood entering a capillary network (capillary bed) is determined by what is happening there at the moment. After lunch, while food is being digested, the blood is circulating freely through the organs of digestion. Between meals, the majority of the same capillary pathways are closed[16] and the blood mostly passes them by, sequestered in the blood vessels (Marieb & Hoehn 2012, Totoro & Derrickson 2013).[17]

---

14 I often insert here the very interesting story of William Harvey and his discovery of the blood circulation to illustrate how difficult it was for early scientists to understand the blood flow and where it came from without the help of microscopes.
15 But not the red blood cells.
16 With the exception of the postcapillary venules, where microcirculatory exchanges also takes place.

The Cardiovascular System

At this point in our considerations, a short "out-breath" can be achieved through a shift in perspective. The teacher presents the students with a riddle. Suppose a German Shepherd jumps a fence and charges straight at you, and you go pale and pass out; or—out of nowhere—you are invited to the prom (or a rodeo, or bowling) by someone you have secretly had a crush on for weeks and your face turns bright red. What is happening in these two instances? You explain that the capillaries in the face are filling up, or losing all the blood they contained, depending on the situation. As the students reflect on this, you may suddenly hear an "OMG" from the back of the room, followed by "that must mean that our inner life has a direct influence on our blood circulation!" Whereby the teacher modestly nods and tries—unsuccessfully—not to blush for joy.

However, another, more skeptical voice may soon join the din and exclaim, "Sure, that sounds great, but we just learned that the capillaries are microscopic in size, are invisible, so how could they make our face red?" This leads us nicely into a further characteristic of the capillaries: They may be unbelievably small, but there are billions of them! In fact, most estimates of the length of all our blood vessels if put end-to-end, would (let the students guess first) not only reach from Chicago to Honolulu, but 2½ times around the earth (~60,000 miles)![18] "OMG" sounds again! With that as background, it is clear that there must be an unbelievably dense network of capillaries beneath the skin (and everywhere else) that enables them to change our facial color so dramatically. Thanks to their extreme smallness—they have an average diameter of 8/1000 mm[19]—they possess a very large surface area in relation to their interior volume. This provides a lot of exchange surface for microcirculation. It is estimated that

---

17   It would be possible to go into much more detail here, describing the different types of capillaries, how the blood flow to capillary beds is regulated by pre-capillary sphincters, etc. However, in my experience, the highest priority at this point in the block is to "keep the blood moving." The students can get stuck if too many details are offered—they "cramp up," so to speak, if the blood does not keep moving sufficiently! Some additional details that relate to microcirculation can be brought effectively later in the block.
18   In the spirit of modesty, however, we must note that not all of these blood vessels are carrying blood at every moment. If we look at how far our 5–6 liters of blood travels every day, it is only about 12,000 miles—which is just back and forth from New York City to LA three times, plus another 350 miles! (George 2020)
19   To give the students some vague sense of how small that is, I often ask them to draw a line on their paper that is one millimeter in length and then imagine slicing it into 1000 tiny sections. Eight of those sections would be the diameter of a capillary.

the capillary network of an average-sized adult provides a surface area of about 1000 sq. miles for exchanges between the blood vessel content and tissues.[20]

Now, as the blood moves out of the capillaries into the network of venules and veins on its way back to the heart, we can try to picture how it has been changing the entire time through its extensive "interfacing" with organs and tissues. After we have eaten, for example, it moves through the intestines and takes up nutrients, after which it goes to the liver, where carbohydrates are drawn out of the blood and a detoxification process takes place. When it passes through the lungs, CO_2 is given off and O_2 taken in; in the brain, large amounts of O_2 and sugar are removed; as it passes through the kidneys, not only waste products but water and salt are withdrawn. In addition, the kidneys frequently secrete a hormone (EPO) into the blood to regulate the production of red blood cells, and another hormone (renin) is given off to regulate blood pressure. As it goes through these and many other organs, the blood is constantly changing—giving and taking—and in this way mediating between and connecting the various organs[21] into a coherent, harmonious whole.

After flowing through the periphery and interacting with organs throughout the body, the blood now returns to the heart through the largest of veins: the superior and inferior vena cava. These vessels bring blood from very different activity centers in the body before releasing it into the powerful rhythms and dynamic of the heart.

The Heart

Arriving at the heart, the central question for the teacher becomes: How do we begin our study of this complex organ without getting bogged down immediately in countless details? I have found it most effective to start with a basic overview of the heart's architecture (anatomy), which provides an elemental vocabulary that we can use as we move ahead. In other words, give

---

20 To make such dimensions more concrete, we imagine, for example, an area of the earth's surface that is 10 miles wide and 100 miles long. (In Chicago we imagine a patch that runs for 10 miles inland from the edge of Lake Michigan to the west and stretches all the way to Milwaukee, Wisconsin, 100 miles to the north). In classes where some of the students speak German, you may even hear the expression "Donnerwetter!" sounding through the rows, when such examples are brought.
21 Much more detail about these organs and their activities will come later in the block.

them a clear spatial orientation first, which means we are going to bring the blood flow almost to a stop for a short time and then try to get it flowing again. Interestingly enough, this is just what the heart does with the blood that comes to it: It puts the brakes on its flow, restrains it for a moment, before sending it vigorously on its way.

The students have already learned the overall pattern of the blood's flow through the heart. They have heard about a right heart from which the blood flows to the lungs (pulmonary circuit) and a left heart from which the blood flows out to the body (systemic circuit). We now look more closely at this. The largest veins in the body, the superior (Ø = 1 inch) and inferior vena cava (Ø = 1.4 inches), flow into the heart like two major rivers that have gathered content from countless tributaries before they reach the end of their journey.[22] But in contrast to major rivers, they do not empty into a quiet, expansive body of water, such as an ocean or a lake. Instead, it is as if they are being awakened out of their dreamy state in a very dramatic and rejuvenating way! More on that soon, but first the heart's anatomy as seen in the drawing below.

The systemic circuit enters the relatively thin-walled right atrium, where its flow is restrained for a moment before passing through the right atrio-ventricular (AV)[23] valve into the thicker-walled right ventricle, where, making a 180 degree

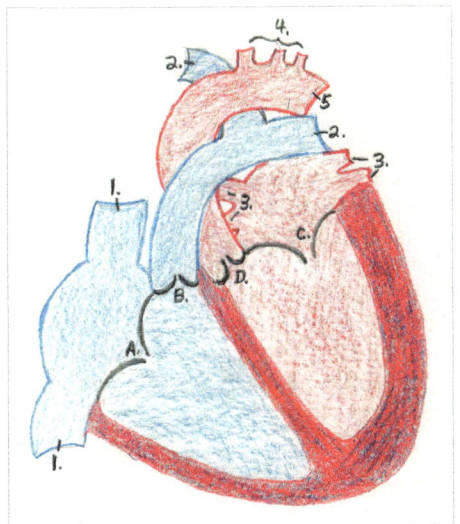

Fig. 2.8 Vertical section showing the blood flow through the human heart, frontal view. 1. Blood flowing from the body into the heart through the superior and inferior vena cava. 2. Blood flowing from the heart to the lungs through the pulmonary arteries. 3. Blood flowing back into the heart from the lungs through the pulmonary veins. 4. Blood flowing out from the aorta into the head and arms. 5. Blood flowing out from the heart down into the body through the descending aorta.

A. Right atrio-ventricular valve. **B.** Pulmonary valve. **C.** Left atrio-ventricular valve. **D.** Aortic valve.

---

22 The coronary sinus from the heart myocardium also opens into the right atrium.

Part II. 10th Grade

turn, it passes out through the pulmonary valve into the pulmonary arteries[24] on its way to the lungs and their capillary beds, where gas exchange occurs, after which the blood flows to the heart via the pulmonary veins that enter the left atrium. From there the blood passes through the left atrio-ventricular (AV) valve into the thick-muscled left ventricle. Here, again, the flow changes direction and leaves through the aortic valve into the aorta (the largest [Ø=1 inch] and most powerful artery in the body) and out into the rest of the body (systemic circuit).

Once we have gained an overview of this straightforward sequence and its anatomical landmarks, we can take a closer look at key characteristics of the heart and its functioning. It is also good to make things a little more difficult for the students (to call on more mental will-activity) by pointing out that the very simple side-by-side illustration of the four chambers is just that: too simple. After noting where the valve plane is on the preceding drawing (frontal view), we show them how it would appear from another angle—from above—if the atria were removed. Having identified the valves on the first sketch, the question becomes: "Which is which" on the new drawing? It would seem clear that the valve farthest to the left (from our perspective) would be the right AV valve, while the valve farthest to the right would be the left AV valve. Fine, but what about the other two?

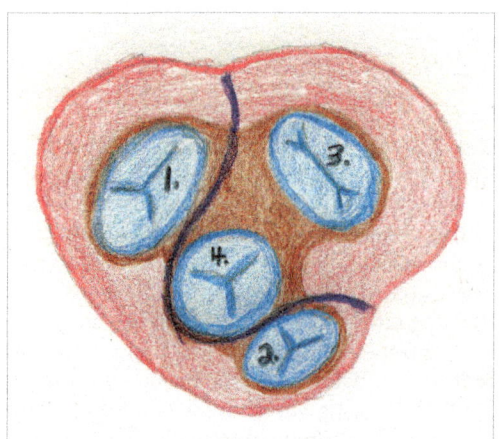

Fig. 2.9 The valve-plane of the heart seen from above. 1. Right atrio-ventricular valve. 2. Pulmonary valve. 3. Left atrio-ventricular valve. 4. Aortic valve. The dark line shows the location of the septum below the valve plane. It separates the right and left ventricle and reveals the spatial relationship of the right and left heart to each other.

---

23 The students find it more transparent—and thus easier to learn—when the teacher uses a term like "atrio-ventricular (AV) valve," which indicates where-from and where-to the blood is flowing, rather than terms like "tricuspid" or "mitral," which are also names for the two AV valves.
24 If it has not been clarified before, they now learn that, per definition, an artery flows away from the heart, a vein toward it, independent of whether the blood is oxygenated or oxygen-poor.

The teacher can then draw a line that shows where the muscular wall (the septum) that divides the ventricles is located and thus show how the pulmonary valve is in front of, and even a bit to the right of, the aortic valve! This is not a simple side-by-side orientation as our first drawing made it appear, but an interesting spatial relationship where the right heart embraces—you might say—the left heart to a degree. If we ask the class to show that relationship with their two hands, they invariably come up with something like: The left heart is like a closed fist that is being, enclosed partially by the right hand (the right heart). This gesture actually indicates something of the nature of the two sides of the heart, as we shall soon see.

A mental picturing exercise that can follow this clarification tells them whether they have really grasped the flow pattern in the heart, or not. With the valve-plane drawing on the board, I ask them to imagine that they are looking down on the valve plane with the atria removed. I ask them to number the valves in the sequence that a portion of blood would move through them and whether the blood at each value is flowing into the board or squirting out into the room (actually, into their faces if they are looking down onto the valve plane).

We can now return to the blood flow in the heart and gain a more real sense for its amazing dynamic. They learn that when the blood flows into the right atrium from the superior and inferior vena cava, the two streams do not collide, but flow past and rotate around each other in a spiraling current. They also learn that such spiraling—vortex-creating—flow patterns exist in the left atrium and in the ventricles during diastolic filling. Medical researchers have even found that optimal vortex formation and flow dynamics are indicators of healthy myocardial activity and play a key role in the diastolic expansion (filling) of the ventricles (Furst 2020).

To help the students develop a more tangible sense for this spiraling dynamic, I often ask them to do the following exercise. Since the blood that is entering the right atrium is creating a clockwise rotating vortex, while the blood entering the left atrium is spiraling in the opposite direction (counter-clockwise), the students can place their index fingers in front of their chests pointing downward, and then rotate the right finger clockwise and the left counterclockwise to emulate how the blood is entering the atria in such a vortex-creating fashion (Holdrege 2002).

If, through such considerations, we are able to give the students a sense of the streaming, rotating, vortices-forming flow of the blood through different areas of the heart, then they will be able to view the amazing patterns found in the muscle fibers of the heart with new eyes. They will experience what a famous New Yorker once called a "déjà vous all over again" (Berra 1974), since many will notice spiraling gestures similar to what we just found in the flowing blood, but this time in the form of continuous sheets of cardiac muscle fibers.

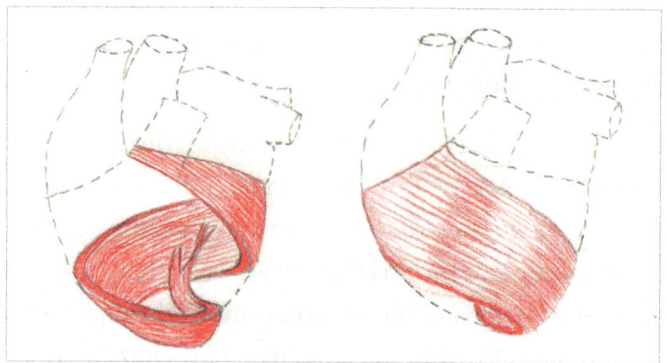

Fig. 2.10 Illustration of spiraling heart muscles.
(After Benninghoff & Goerttler 1980)

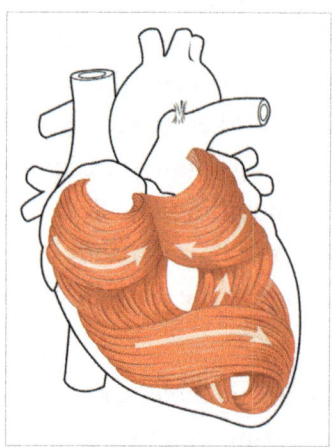

Fig. 2.11 Illustration of the complex and elegant muscle patterns that swirl and spiral around the chambers of the heart. They form a figure-eight pattern around the atria. Deeper ventricular muscles also form a figure-eight around the two ventricles.
(Source: Biga, Dawson, et al. Anatomy & Physiology. Creative Commons Attribution-ShareAlike 4.0 International License)

As I mentioned, many of the students will be struck by the amazing parallels between the flow of the blood and the multiple vortices of the heart muscle. Do we have a chicken and egg question here, similar to the one we started our study of the CVS with? Is there a straightforward cause-effect relationship between the spiraling muscle fibers and the spiraling blood flow?

The Cardiovascular System

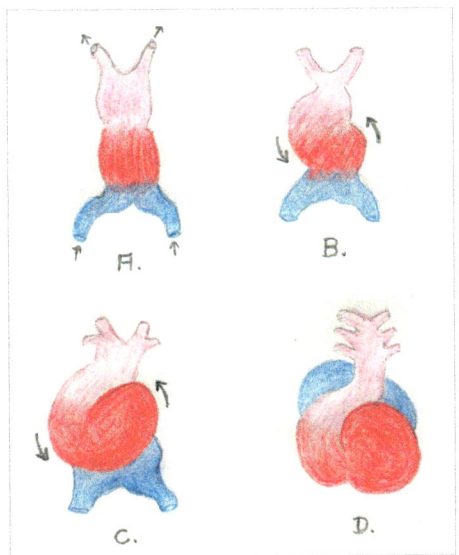

Fig. 2.12 Frontal view of the developing human heart during the fourth and fifth weeks. Ventricle pole in red, atrial pole in blue. **A.** 22 days: a simple heart tube has formed. **B.** 23 days: twisting and looping begins. **C.** 24 days: twisting and looping continue. **D.** 35 days: the ventricle and atrial poles have reversed their location, with the atria above (superior) and the ventricles below (inferior).

The embryological formation of the heart can help us answer this question. After passing through an early tubular phase (22 days) with the blood flowing directly through it, the embryonic heart is gripped by spiral-forming tendencies and begins to loop and fold. Within the next two weeks, these movements are complete and the atria and ventricles of the future heart have assumed their final adult positions.[25] During this looping phase, two distinct currents (the pulmonary and the aortic) have formed and flow through the heart. They stream and loop by each other without mixing. In the "quiet zone" between these two currents, the septum that will separate the two halves of the heart forms and will develop further in the next 10 days. What does this process show us? It shows how blood flow and organ formation are mutually engaged in this process. While, on the one hand, the looping of the heart tube causes changes in the blood flow, on the other, the way in which that blood flow constitutes

---

25   In some classes I have found it worthwhile to do a simple re-creation of this process using hand gestures. I ask the students to place their left hand in front of them, fingers pointing upward, palm outward. Below this the right hand is placed in the same orientation with the fingers just touching the base of the hand above. Then both hands begin to move simultaneously: The upper hand rotates leftward (counterclockwise seen from below) and downward, moving into the horizontal, while the right hand rotates rightward and upward until the two cupped palms are facing each other. The left hand now represents the inside (base) of the ventricles, the right hand the inside (roof) of the atria. If we let the one contract while the other opens in alternation and then move the hands "gently" over the location of the heart (right hand above), we have created a rudimentary experience of the beating heart.

Part II. 10th Grade

itself (independent, looping currents) influences how the looped heart form will divide into two separate halves. Through the looping and the formation of the septum between the two blood currents, we see how form influences flow, and flow influences form: These two polar aspects of the circulatory system are interwoven and interdependent (Benninghoff-Goerttler 1975, Holdrege 2002, Sadler 2000, Tortoro & Derrickson 2013).

In what Branko Furst (2020) calls an "opus magnum" on intracardiac blood flow, Professor Ares Pasipoularides, of the Duke University School of Medicine, concludes that form and function in the heart have become one: "The spiraling and looping pattern of myocardial fibers, including the vortex cordis, is a reflection of intracardiac blood movements. In the heart, form and movement unite in a rhythmical process in space and time: The organ, a form in space, is simultaneously a movement in time… the movement of the cardiac walls is a physical replication of the creative fluid flow movements which they enclose (Pasipoularides 2010, pp. 301–302).[26]

So what does the basic movement of the cardiac muscle look like when it alternates rhythmically between the diastolic and systolic phases of the cardiac cycle?

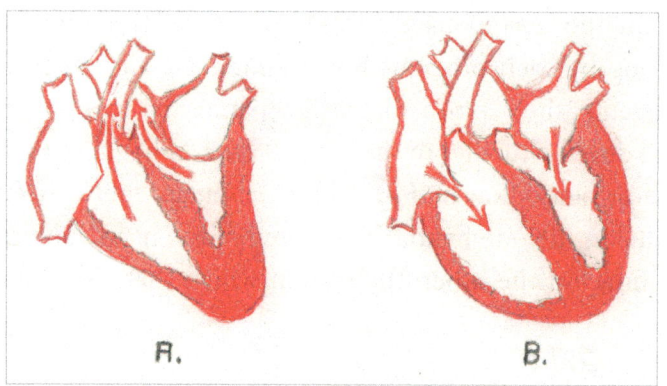

Fig. 2.13 The cardiac cycle. **A**. systole. **B**. diastole.

---

26 Pasipoularides also notes: "In a way, what the blood does as a fluid has fashioned muscular histoarchitectonics of the cardiac ventricles. In spiraling paths, myofibers sway down to the heart's apex and then rise again to its base. They make the same movements and emphasize the same vertical streaming of the filling vortex within the ventricles. Therefore, the intraventricular blood flow and the pattern of the muscular histoarchitectonics of the heart are mutually intertwined" (2010, pp. 301–302).

During the systolic phase[27]—when the ventricles contract, the AV valves close, and the aortic and pulmonary valves open—we have a dramatic twisting of the apex of the heart in a counterclockwise direction as it moves downward and pulls the valve plane with it. During the diastolic phase, the ventricle unwinds clockwise, the valve plane rises, and the ventricle opens up to receive the blood that is spiraling into it (Furst 2020).[28]

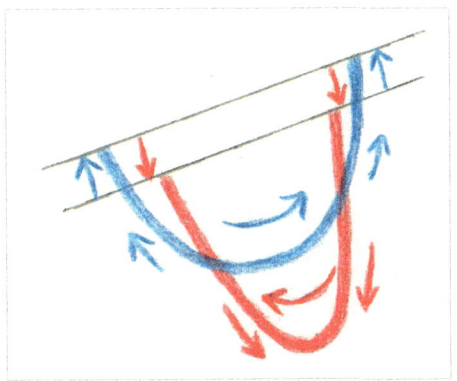

Fig. 2.14 Sketch of cardiac cycle showing the counter-clockwise descending & squeezing gesture of the ventricles in the systolic phase, which contrasts with their clockwise ascending & opening gesture during the diastolic phase. The two, almost horizontal lines at the top of the sketch indicate the rising and falling of valve plane during this process, thereby accentuating the receptive gesture of the heart as it rises and opens to receive the incoming blood during diastole.

A simple sketch such as the one above can give a sense for the process just described. It also helps to have the students mimic this dynamic with their hands. Starting with a hand opening upward, they then pull it downward with a counterclockwise twisting motion, then untwist it again as it moves upward and opens receptively. Students often think of milking a cow after making these movements! From there, I often ask them to do something similar with their whole body. Raising their arms upward in a bowl-like, receptive "ah" gesture, and then twisting downward (counterclockwise) ~180 degrees into a crouch, before twisting upward again and opening to receive what comes to them from above. We then talk about how extreme the polarity is between these two movements, and concur that it is truly amazing that the heart can make such a double-gesture

---

27 The terms *systole* and *diastole* will be used here in reference to the ventricles only. It is possible, of course, to speak of an atrial diastole, which takes place at the same time as the ventricular systole, and of an atrial systole that parallels the ventricular diastole. I have found, however, that this tends to create confusion for the students during discussions, and hence, I prefer to refer to the atria merely in terms of relaxing (expanding) and contracting.

28 If the teacher feels s/he can complicate the picture a bit more without losing the students, it can be noted that when the ventricle is twisting counterclockwise in the systolic phase, the atrium above is rotating clockwise and relaxing, whereas when the ventricle relaxes upward in a clockwise direction during the diastolic phase, the atrium above is contracting counterclockwise.

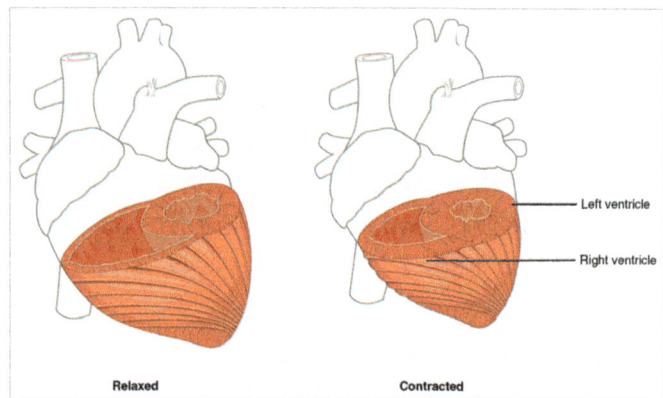

Fig. 2.15 Muscle thickness in the right and left ventricles. The myocardium in the left ventricle is significantly thicker than that of the right ventricle. The ventricles are shown in both their relaxed and contracted states. Also visible is the difference in the relative size of the two lumen.
(Source: Biga, Dawson, et al. Anatomy & Physiology. Creative Commons Attribution-ShareAlike 4.0 International License)

roughly every second for an entire lifetime—well over two billion times if you live to be 70. And this without ever going on vacation or taking the weekend off!

This remarkable phenomenon can be amplified even more by describing the amazingly precarious situation that the left ventricle finds itself in during this process. The left ventricle is much thicker and has considerably more muscle mass than the right one, which means it requires a much larger oxygen supply in order to function.

However, as it turns out, the extravascular compression of the left ventricular muscle is so powerful during the systolic phase that the arteries in the inner layers of that muscle (the subendocardial coronary vessels) are squeezed so tightly that they are "pressed white," and at times the blood in them is even forced to flow backward![29] As a consequence, these inner layers only receive full

---

29 A phenomenon that is otherwise unheard of.
30 "Extravascular compression during systole markedly affects coronary flow; therefore, most of the coronary flow occurs during diastole. Because of extravascular compression, the endocardium is more susceptible to ischemia (insufficient blood flow to provide adequate oxygenation), especially at lower perfusion pressures. Furthermore, with tachycardia there is relatively less time available for coronary flow during diastole to occur–this is particularly significant in patients with coronary artery disease where coronary flow reserve (maximal flow capacity) is reduced" (Klobunde 2017).

circulation (perfusion) for two-tenths of a second per heart beat.[30] This means that these layers are fully circulated for one-fifth of a human lifetime (Klobunde 2017, Schoeffler,1975). This is a one-of-a-kind situation and raises the question: How can the left ventricle—and with it the heart as such—survive under such conditions?[31]

The magic word here is "rhythm." In the following quote, Husemann and Wolff describe the heart's rhythmical activity as the secret behind its lifelong activity and behind health in general:

> *Systole corresponds to the contracting dynamic of the nerve-sensory system, diastole to the expanding tendency of the metabolic system. The heart's function alternates between them. It is decisive that the heart is completely and entirely opened first to one and then to the other impulse. Thus, maximal work alternates with absolute rest. At the same time this signifies a maximal amplitude of frequency. The refractory state (diastole) corresponds to extremely deep sleep, the always maximal contraction to intense work.*
>
> *Here is the secret of health and the seeming indefatigability of the heart. It is not ceaselessly active without rest, but rather, after every activity an absolute rest is necessarily inserted at the optimal moment. It is only for that reason that the subsequent performance is possible. Not indefatigability, but rather guided rest alternating with intensive activity, that is, a strict rhythm makes lifelong function possible and thus shows the archetypical picture of health.* (1987, pp. 361–362)

From this perspective we can lead the students in a discussion about the role rhythms play in other aspects of our lives. Rhythms like sleeping/waking, summer/winter, day/night, main lesson/recess, and school year/vacations will usually come up and can be explored further as to their significance. As the block progresses we will be able to refer to this very important theme in numerous contexts.

---

31 As it is, the vast majority of all acute myocardial infarctions involve the left ventricle.

Peripheral Circulation

With the contraction of the left ventricle, the blood is ejected out into the aorta (the largest artery in the body) in a vortex-like spiral pattern and with a dynamic that we compared earlier to a mountain stream. Indeed, the flow is swift and oxygen-rich as it races out into the various parts of the body, but it also has a characteristic that no mountain stream ever has—a pulse. The blood that leaves the heart is the fastest flowing in the body, but slow compared to the pressure wave (the pulse wave) that shoots down through the aorta and the arteries that branch out from it at up to 10 times the speed of the flowing blood itself. Whereas a pressure wave can move from the aorta to the foot in about 0.2 seconds, the blood that left the heart at the same time will need roughly 2 seconds to get there (Brettschneider 2002).

The distinction between the flow of blood and the pressure wave is an important one because they are often viewed as the same thing. One easy way to help the students understand this is to point out that when waves pound the shore of a lake on windy days, it does not mean that all the water is flowing out of the lake and onto the surrounding land. No, only the waves are moving through the water at high speeds. The surfers are not swept to the shore as they wait for the next big wave, but only bob up and down each time one passes by. Another example could be the sound waves we discussed in the 9th grade. The air is not blowing toward our ears when we hear a sound, but rather, sound waves are passing through it. Another significant difference between the two is that while flow velocity decreases quite rapidly as it moves further from the aorta, the speed of the pressure wave increases steadily.

So, although we have received a boost in the blood flow—only a boost, since it was already flowing when it entered the heart—we now have something quite new and significant: rhythm! The pressure wave brings rhythm into the circulating blood. Something of the polarity between systole and diastole that we discussed earlier is thus passed on to the rest of the organism. Rhythm is not just another added characteristic when it appears, but, as the students will know from the world of music that is so central in their lives—it has a very tangible harmonizing and organizing influence anywhere that it appears.

It is also important for the students to understand the distinction between an unchanging beat and a rhythmical process in this context. Otto Wolff provides a helpful characterization of that difference.

The beat repeats, rhythm renews. The beat is the persistent repetition of the same event; rhythm is the continuous modification of a preceding similar element. Thus, there is a harmony in rhythmical events that leaves open new possibilities of development. Beat, however, through the repetition of the same, may well be more exact, but it is also more rigid and fixed. Here an event is led over into a state that remains the same and is held there. Beat leads to isolation from interconnection; it creates a separate independent life, whereas rhythm is an expression of an active harmony. Beat is dead rhythm. It is for this reason that beat is so tiring. True rhythm, however, is vitalizing. Thus, beat is suitable to mechanics and rhythm to living systems. (1987, pp. 320–321)

At this point, we can ask the important question: What determines the heart rate in the first place? The students all know if they run around the school building or do 50 push-ups their heart rate will increase. They all know that their muscles need more oxygen for this activity and thus they need the oxygen-carrying blood to arrive more quickly. That is quite straightforward. Now we ask them what they think would happen if we had them relax in a comfortable arm chair and then— through artificial means—caused their heart rate to increase. We explain how this can be done with a so-called "pacemaker." It may seem an odd thing to do, they comment, but scientists often do out-of-the-ordinary things to learn more about the ordinary.

What's the result? The students, for the most part, think that it will lead to an uncomfortable experience of trying to relax but with blood rushing through one's body at a very high rate. They picture it as being similar to turning up the speed of a pump—the faster it goes, the faster it flows.

But alas, that is not what happens. Studies of this nature have shown that when a pacemaker induces a rapid heartbeat (pacemaker tachycardia) up to four times the normal rate, the strength of the heart contractions increases, as does the aortic pressure, but the volume of blood that flows out from the heart each minute (cardiac output) does not change! (Furst 2020, Lauboeck 2002)

First of all, we ponder how this could be possible. Discussion leads to the conclusion that with each heart beat less blood must be ejected, which means that there must be less blood returning to the heart under these circumstances, which means the heart rate as such is not causing the blood to circulate faster, is

Part II. 10th Grade

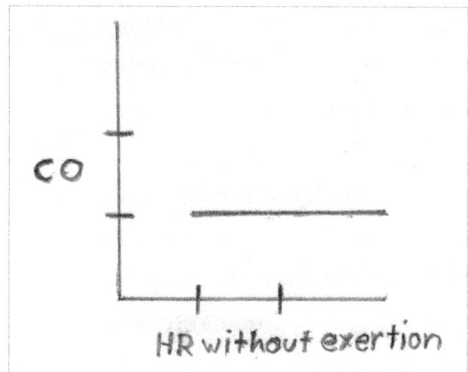

Fig. 2.16 Without exertion, cardiac output (CO) remains the same despite a pacemaker-caused increase in heart rate (HR). (After Lauboeck 2002)

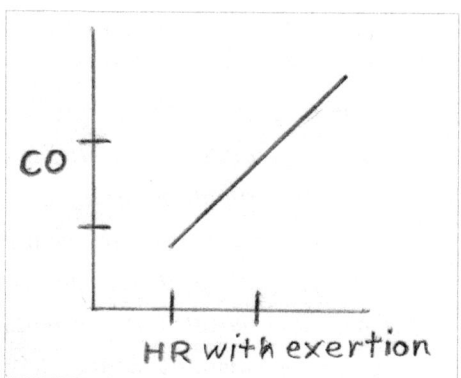

Fig. 2.17 During exertion, cardiac output (CO) increases as the heart rate (HR) increases. (After Lauboeck 2002)

not driving the blood circulation as a whole. This relationship can be illustrated with a graph (Fig. 2.16).

Returning to normal life, we find—not surprisingly—that when someone runs quickly their heart will beat approximately three times faster than when they are at rest, and five times as much blood will pass through it (cardiac output) and the body. See Fig. 2.17 above.

No surprises here: The more we exert ourselves, the more rapidly the heart beats and larger quantities of blood flow out to the body. Why is that? Almost everyone in the class will (should) know immediately: the need for oxygen. Although in everyday life overall flow of the blood involves more than simply meeting oxygen needs, in the case of strenuous activity, it is certainly primary. As it turns out, when the data is put together and graphed, it looks like this.

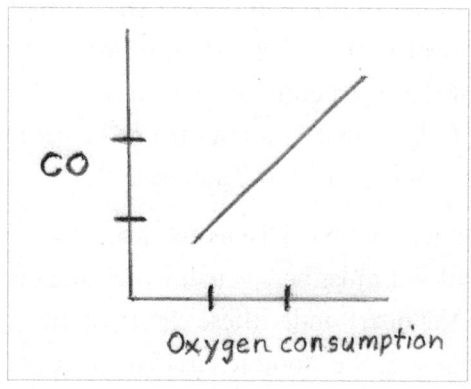

Fig. 2.18 Cardiac output (CO) increases in correspondence with increased oxygen consumption. (After Lauboeck 2002)

The Cardiovascular System

The 45-degree angle on the graph tells us that cardiac output and oxygen consumption are perfectly correlated: when oxygen needs goes up, so too do the HR and CO, when oxygen needs are lower, the HR and CO goes down. It is also known that when the body temperature falls below normal, metabolic activity and oxygen consumption decreases. By a body temperature of 89.6 degrees, the metabolic rate drops by 50% and cardiac output decreases by the same percentage (Lauboeck 2002).

Looking at all this data, it gradually becomes clear that the ultimate determinate of blood flow is tissue-metabolism. Every organ system in the body has its own specific capillary type—from tightly-closed ones in the brain to the porous walls of the kidneys. As we learned earlier, the relative amount of blood entering an organ's capillary network depends on what is going on there at the moment. Body tissues that have high metabolic requirements have extensive capillary networks. When an organ is active, the entire network fills with blood. During exertion, for example, the volume of blood that flows through the muscles can increase 15–20 times, while concurrent with that, the flow to the kidneys and liver decreases. Only the flow to the brain remains constant.

This interaction between the blood and the bodily tissues is referred to as microcirculation. One central riddle for physiologists has always been: How are the metabolic needs of the tissues and the cardio-respiratory response thereto mediated? It was not until the 1990s that it was discovered how the red blood cells not only transport oxygen to the tissues, but actively regulate the supply and demand of tissue oxygen.[32] Not only do they play an essential role in increasing the diameter of the blood vessels (vasodilation) to increase the blood flow to areas that are oxygen depleted, but they do this in such a way that the vasodilation progresses against the blood flow—upstream—to "recruit" more arterioles leading to the capillary beds in question, thereby increasing the flow to them even more! At the other extreme, there are times when all capillary flow in an area ceases. This happens, not to the surprise of the students (the teacher hopes), when local metabolic processes have no "needs" at the moment (Brettschneider 2002, Furst 2020).

---

32 The blood as such, and the red blood cells in particular, will be focused on later in the block.

This phenomenon is an important stepping stone in helping the students realize that the blood is not, in the words of Branko Furst (2020), "an inert fluid 'pumped' around the circuit by the heart, but a 'self-moving' agent with flow directly coupled to the metabolic needs" of the organism. We will return to this topic shortly, but first a description of the all-permeating presence of "microcirculation and flowback" in the human body.

Microcirculation and Flowback

The tissues of the body are constantly producing fluids and substances out of what they receive from the blood. They also give back into the blood what they have produced, as well as the waste products there from. This exchange between bodily tissues and the blood requires a mediator: the interstitial fluid. This fluid fills the spaces (extracellular spaces) between the cells of a body tissue. It comes from the blood plasma, which has left the capillaries and circulates around the cells bringing to them water, nutrients, oxygen and other solutes. The solutes that leave the cells enter the interstitial fluid and then enter the capillaries. All together, the extracellular spaces contain around 10.5 liters of interstitial fluid, which is more than three times the amount of fluid blood plasma (which is around 3 liters in an average-size human being).[33] The amazing thing about this exchange is that approximately 200 liters of fluid leave the blood vessels and return changed every day.[34] This means that the total fluid volume of the blood plasma leaves and returns (flowback) to the bloodstream about 70 times per day. It is also important to note that these microcirculatory processes are highly independent of the cardiac cycle (Brettschneider 2002, Lauboeck 2002, Marieb & Hoehn 2012).

How is all this activity possible without being coupled to the pressure and flow coming from the heart? One key factor is that we are dealing with a dimension of existence very different from the world around us and from the fist-sized human heart. We learned already (2.3.3) how incredibly small the capillaries are. We now need to discuss with the students a central consequence of that smallness, one that "changes the playing field" when it comes to circulation and flow.

---

[33] The cells themselves, in sum, contain about 30 liters fluid.
[34] The preponderance of this takes place in the kidneys.

I like to introduce this aspect of our topic with a few examples. One that some of the students enjoy is when the teacher stands in front of the class and holds a smallish piece of chalk over their head and lets it drop, followed by the question: Why did it fall? Easy questions yield a rapid response: Gravity. Then the teacher claps two dusty erasers together. A cloud billows up, some chalk dust settles to the floor, but much of it drifts around for a time, and if the teacher fans it, it will drift in this direction or that, depending on the impulse given. What caused the different result in these two demonstrations? We are, after all, dealing with the same substance, in the same room, at roughly the same time and temperature. Well, one is dust and the other is a solid chunk. We know how dust sails around, chunks don't—but why? Size! The smaller things get, the larger the ratio of their surface area to their volume (SA:V). The larger the surface area relative to the volume of the interior, the less effect gravity has on it. Chalk dust consists of tiny chalk particles that have a much larger SA:V ratio than a large piece, which frees them to a large extent from gravity's pull.

Life is actually full of examples that show this. I like to tell the students how I noticed one day—back in the romantic 1980s when I lived in Vienna, Austria—that my bicycle sitting in front of the house was covered in a rust-colored dust. A few days later it hit me: "OMG!" that dust comes from the explosion of Mount St. Helens in the state of Washington, USA—5500 miles away! The volcano had erupted a few weeks prior. What had been huge masses of hard volcanic rock that formed a mountain towering above the Pacific Ocean was suddenly blown into tiny pieces, some of which were carried around the earth to where I lived (and much further) and provided my garden—finally—with some good American soil! I tell them how, for more than a month after the explosion, people in Portland, Oregon (more than 70 miles distant) had to wear face masks to keep from inhaling the volcanic dust particles that were still floating in the air (some of which were less than 5 microns in diameter and smaller than a red blood cell).

The teacher can follow such a story with numerous other phenomena that relate to SA:V, things that the students are familiar with. One possibility, for example, is to ask where clouds come from, how they manage to float around as they do, and why rain drops suddenly fall out of them?[35] In short, our "playing

---

35   They will know the answer to these questions, but not with the emphasis on SA:V.

field" in the world of microcirculation is not only small, but gravity-free. That is a huge difference! Movement is not the same issue any longer. This can be illustrated by John Ruskin's famous comment that although Newton explained to us what caused the apple to fall from the tree, he failed to seek the answer to the much more difficult question: How did the apple get up there in the first place (Ruskin 2012)?

So how does the apple get up to the top of the tree? Although not a simple matter, one key aspect is that the xylem cells through which water is transported up a tree (some trees can reach a height of 200–300 feet!) typically have a diameter of only a few microns. "Small is beautiful!" (Goodbye gravity!)

Certainly many factors—diffusion, osmosis, etc.—are involved in microcirculatory processes, but it is pivotal for all activity at this level that the force of gravity no longer plays the same role that it does when things get larger. In his extensive discussion of the significance of smallness in the life of organisms, Otto Wolff speaks of "primary streaming" as a phenomenon that one can observe at many levels in an organism. He points, for example, to animals such as mollusks, where their circulation takes place without a heart. In particular, he notes how the lancelet (*Branchiostoma*)—a creature that the students will learn much more about in the 11th grade—has a circulatory system where the vessels all remain capillary size and hence no heart is needed. Also of great interest for the 11th grade block is Wolff's conclusion that the reason organisms have cells as their basic units has to do with size: "The cell's smallness is a pre-condition for its becoming a carrier and element of life. The large surface area made possible by the cells' smallness is indeed the very reason for its existence" (Wolff 1987, p. 347).

But all is not small in the human being and in many other creatures, so when the blood vessels get larger gravity becomes, once again, a factor. As the blood flows back toward the heart in the gathering and ever-larger-growing veins, we enter what we called earlier the mellow flowing, high volume, "big river" phase of the blood circulation. No pressure waves, no pulse is active here. What now?

Sensing and Harmonizing

As Wolff points out, as the heart opens up to the spiraling bloodstreams coming from above and below (superior and inferior vena cava), it not only receives, but also perceives. The inflowing blood (venous return) causes the right atrium to expand. In the walls of this atrium are so-called low-pressure baroreceptors (atrial-stretch receptors) that sense the volume ("fullness") of the returning blood. When there is an increase in blood volume, for example, it is regulated by the baroreceptors and the heart's response influences the situation in three primary ways.[36] First, atriopeptin (atrial natriuretic peptide, ANP)—a protein hormone—is secreted by the atria. By secreting this hormone the heart is able to influence blood volume very significantly. ANP causes the kidneys to increase the secretion of salt and water, which has a direct effect on the fluid volume of the blood. Secondly, the secretion of ANP causes the peripheral blood vessels to widen (vasodilation) so that there is less resistance to blood flow. We see how the perception of increased venous return is a signal to the heart that lots is going on in the periphery. High levels of blood flowing back to the heart is like a cry for help—it means that lots of O_2 is being consumed by the tissues and that they need more, ASAP! So, thirdly, the heart responds directly: it begins to beat more rapidly[37]—three to four times more rapidly, if necessary—in order to provide the O_2 needed[38] (Constanzo 2013, Marieb & Hoehn 2012).

We see a similar sensing and then harmonizing gesture in the heart's relationship to breathing. Chemoreceptors are located in the blood vessels that exit the heart (the aorta and the carotid arteries) and that carry freshly oxygenated blood from the heart into the body. Through these receptors it is possible to sense whether the blood that was sent to the lungs from the right heart (pulmonary circuit) and which is now leaving the left heart to return to the body (systemic circuit) has taken up sufficient O_2 and eliminated enough CO_2 to optimally serve the tissue needs downstream. If O_2 concentrations are low, or if arterial pH has fallen, then the lungs are immediately stimulated—

---

36 One could go into more detail here, but it is always a question of how many specifics are possible before the students become overwhelmed and the flow dries up.
37 This is known as the Bainbridge reflex.
38 This should be no surprise to the students based on the correlation we saw earlier between cardiac output and O_2 consumption.

already within one or two breaths—to increase the rate and depth of breathing (hyperventilation), which increases O_2 levels, and through the exhalation of CO_2 causes blood pH to return to normal. While these are the most palpable influences of such chemoreceptors in everyday life, there is a wide range of less noticeable conditions in the blood—including levels of potassium, glucose, insulin, and immune-related cytokines—that chemoreceptors also respond to (Eckert & Butler 2017).

These examples can prepare the class for a discussion about the central role that the sensing and harmonizing activity of the heart plays in the health of the entire organism. It has become clear that the heart has two central capacities that make it indispensible for a healthy organism. With the help of the blood vessels leading to and from it, the heart is, first of all, receptive to and perceptive of what flows into it from all corners of the organism and is able to "diagnose" or "read" from the flow and the make-up of the blood what is needed to support and balance out the activity of the other bodily systems working in the periphery. On this basis it is then able to respond to those needs itself (change in heart rate) or stimulate responses elsewhere (in the kidneys and lungs, for example) in order to harmonize again what has—for the moment—lost its equilibrium.

Polarities and Rhythm

Having determined that the two poles of this process involve being a) "receptive and perceptive" on the one hand, and b) responsive with a new balancing impulse on the other, we can step back and look—with the help of these two concepts—at what we have learned about the CVS so far in this block.

Starting with the powerful, twisting dynamic of the left heart and with everyone doing the two hand exercise we practiced earlier, we see how during the systolic phase the atrium is receptive and perceptive (through the atrial stretch receptors) as it opens upward (clockwise)[39] while, concurrently, the ventricle is responding by sending out a powerful pressure and flow impulse into the periphery as it twists downward (counter-clockwise). Then the process is reversed: as they seem to reunite, with the atrium contracting (counter-clockwise) and the ventricle unwinding upward (clockwise) to receive the blood

---

39   When viewed from below.

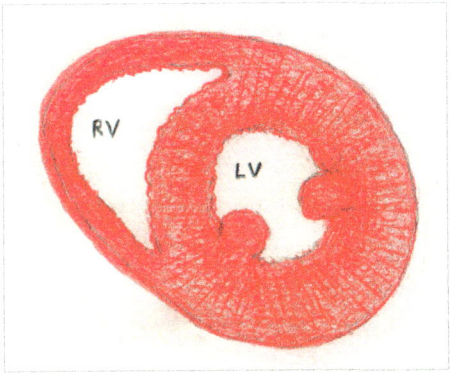

Fig. 2.19 Transverse section of the human heart (inferior view) showing the difference in thickness and shape between the right and left ventricular walls. (After Textbook of Cardiology.org)

from the atrium. We see how both poles of the left heart (the atrium and the ventricle) have a receptive phase and a contracting-ejecting one, and that they move between these extremes every second of every day. By creating an ongoing rhythm the heart itself balances out these extremes—which, for the inner layers of the ventricle, as we saw earlier, is a never-ceasing life or death issue.

If we move to the right heart, we see the same polarity, but the atrium seems to be primary in that it receives flow from the whole body, whereas the right ventricle is a thinner and much less powerful muscle than the left ventricle.

So if we compare the two sides of the heart, the students will usually concur that the left heart is more ventricle (impulse-sending) in nature and the right heart more atrial (receptive, perceiving) in its emphasis.

This is supported by the fact that on average the diastolic filling-phase in the atria is over 1.5 times longer in the right heart than in the left (0.7 sec. vs. 0.45 sec.), whereas the ventricle's systolic pressure-phase is three times longer in the left heart than in the right (0.35 sec vs. 0.1 sec)(K. Llewellyn 1994).

We can step back even further and compare the venous and arterial sides of the circulation from this angle, and—lo and behold—we see a similar polarity manifesting there as well. As we learned earlier, the rapid, pressure-wave-permeated "whitewater" passes through thick-walled elastic/muscular arteries that hold their shape and round lumen very consistently, whereas the thin-walled veins are very receptive to changes in blood flow, relaxing and expanding when blood volume increases, or allowing themselves to be pressed flat when the muscles around them contract. In short, the veins are receptive to and are influenced strongly by their surroundings, while the arteries are more self-contained and resist outer influence through their elastic muscularity.

It is not surprising then—as we heard—that only 15% of the blood volume is to be found in the high pressure arterial system, whereas the vast majority of the blood spends its time meandering through—in a much more leisurely, tissue-interactive way—the low pressure areas. Seen from our current perspective, the venous system clearly shows the more receptive atrial gesture, while the arteries are more similar to the muscular, impulse-generating ventricles.

And if we can step back one more time (before exhaustion sets in!) and compare the heart to the capillary system, we see at one extreme the permeable, periphery-open, pulse-free, receptive nature of the capillaries—where even the red blood cells within them are sensitive to the metabolic needs of the tissues—and at the other end of spectrum, a powerful, muscular, pressure-wave-creating heart, that beats without pause.[40] Mediating between these two extremes are muscular arteries that carry the blood from the ventricles, on the one hand, and the atrium-linked veins, on the other.

In discussion we might then come to the characterization that in the peripheral circuit we have an expansive diastolic gesture, in the heart a muscular systolic one. In the former we are freed from gravity and find primal flow that needs no outer "push"; in the large vessels of the heart region, where gravity rules again, the blood flow is given a new pressure-creating "whitewater" impulse.

Looking back at this point with the class, we can ask: What have our studies of the heart and circulation shown us so far? One key aspect is that everywhere we find polarities, and these polarities are mediated by a rigorous and rhythmical alternation between extremes. We spoke several days prior about the central role that rhythms play in our lives (sleeping/waking; work/play) and in nature (day/night, summer/winter). It should thus be no surprise to find that rhythms are so essential in our own physiology. We can also expect to find rhythms playing a fundamental role in the life of other organs and processes that are still to come in this block.

---

40   Which happens, to say it one more time, in response to the needs of the periphery.

Inner and Outer

Around this point in the block, we can remind the students of the discussion we had on the first day, when we asked if a list of all the organs and processes in the human body would give us the whole human being. The clear response then was "no," there is a whole other realm—not perceptible to the outer senses—that we call our inner life: thoughts, feelings, memories, sense perceptions. The teacher can now raise the question whether there might be—even though the inner and outer levels of our being are so different—commonalities between these two realms? Do they share any characteristics? Can we also find something like the systole-diastole rhythm in our inner life? The student's reflections will go in different directions, but if they do not come to it on their own, the teacher can ask them for some examples of characteristic feelings, and write them on the board. After a few are listed, we then see if we can make gestures that correspond to each. Happy: open arms, smiling, bouncy. Sad: heavy, withdrawn, turned inward, and so on. Before long, the students will notice that each feeling can usually be paired with one that is its opposite, and that the corresponding gestures are also the converse of each other. We now make a whole list of such "feeling-pairs" on the board. It soon becomes very evident that the one pole is open, expanding, the other closed, withdrawing. We can point out that psychologists often use the general terms "sympathy" and "antipathy" (yes and no) for these two tendencies. Returning to our question, it is easy to see that our feeling life moves between two poles that have the similar qualities to the systole/diastole dynamic we found in the heart.

We all know, too, how our own heart rate can be influenced by our feelings. The students will all agree that when they are standing on stage behind the big curtain, ready to step out before an auditorium full of students and parents, their heart is beating wildly. They also agree that if their favorite pop group, for example "One Direction (1D)," were to suddenly step into the classroom, some of them would feel that their heart was going to stop (shock!)—before it took off beating at a wild, unstoppable pace (excitement!).

Since students at this age seem to enjoy most those teacher-anecdotes that show the teacher at less than their very best, I often share with them how, on my first teenage date, as I was sitting in the movie theater trying to get up the courage to put my arm around the young lady next to me, my heart was racing so rapidly I was afraid she might be able to hear it! But worse yet, when I finally

Part II. 10th Grade

made the attempt, I hit her on the side of the head with my elbow, at which point my heart almost stopped. Not to mention that my face glowed red as a tomato, while hers turned pale as a parsnip.[41] (Needless to say, the rest of the movie passed before my eyes without being noticed.)

Many studies have been done identifying the relationship between emotions and heart activity. Our heart responds to feelings like anger, frustration and anxiety, for example, with more erratic patterns—which also makes it harder for us to think clearly. On the other hand, emotions like compassion, appreciation, and caring produce much more harmonious heart rhythms, as the graphs below show (McCraty 2015, Cacioppo 2000).

After exploring these interesting parallels between our inner life and our bodily one, we can look at the way we use the word "heart" in the English language. I like to give the students about five minutes so each can make their own list (no sharing!) of terms and phrases that include the word heart.[42] Most

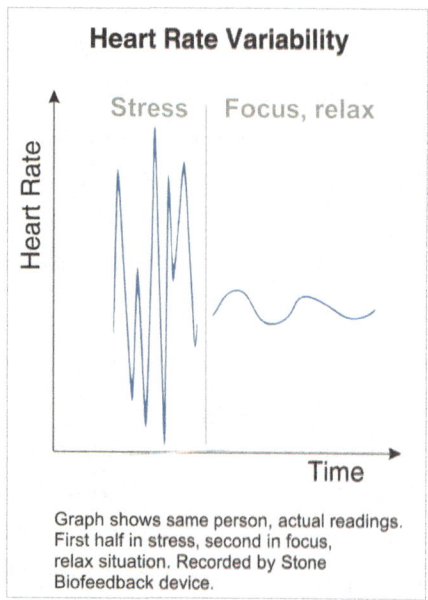

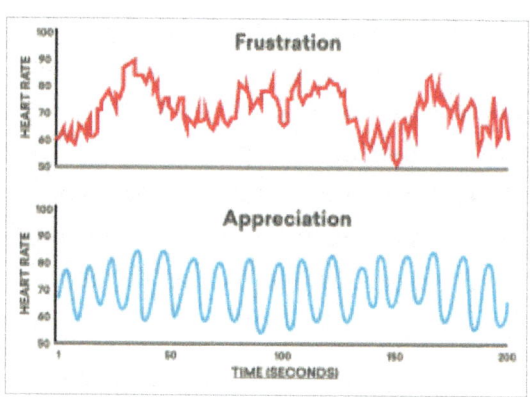

Fig. 2.20 Emotions influence heart rhythms.
(A. Source: Marek Jacenko thebiofeedback.com. Creative Commons Attribution-Share Alike 4.0 International license. B. Modified after Holdrege 2002)

---

41 Being the alert students that they are, the class will inevitably point out, of course, that blushing and turning pale are related more to the capillary network (the peripheral circulation) than to the heart!

42 The teacher can also point out that we would come to a very similar result if we used another language. In fact, if there are exchange students in the class, they can be asked to do it in their own language to confirm this (*Coeur, Corazón, Herz*...).

of them can come up with 10 or 15 very quickly and easily: warm-hearted, cold-hearted, faint-hearted, heart in your mouth, heart of gold, broken-hearted, heartless, give your heart away, heart-to-heart, heartfelt, light-hearted, whole-hearted, close to your heart, and many more. Our list quickly fills the board and we could go on.

A discussion can then follow about what these expressions have in common.[43] It is fascinating how actively some classes will take up this aspect of our "inner life" in discussion and enter into serious reflections about what it means to be "truly human" (or "inhuman").

Warmth

Returning to the biological (but with the psychological perspective still accompanying us), the students are often surprised to learn about the pivotal role that the heart plays in the creation and regulation of our warmth organization, and how much our waking consciousness and sense of well-being depends on this. This can be seen very clearly through the changes that take place during hypothermia—the artificial sinking of the body temperature that is necessary during difficult and long-lasting surgeries. It is necessary because it decreases the O_2 needs of organs, and sinks blood pressure, heart rate and cardiac output. Already the drop from 98.6°F to around 97° is experienced as very uncomfortable and the patient begins to shiver. At 93–95° the patient loses the control they normally have over their feeling life. With a stiffening of the muscles at about 92°, coordination of the body becomes difficult and speech is often slurred or indistinct mumbling. At 87–90° a patient can still follow what is being said to them, but they will not be able to remember it later. At 86° consciousness is lost altogether (Kranich 2003).

The fact that most of us have never experienced these conditions tells us how finely regulated our warmth organization is. Most students will have had some experience with fever and know how discombobulated they can become: shivering, shaking, chills, aching muscles and joints, weakness, loss of appetite, sweating, mental activity becomes overwhelming. In the United States a reflex-like negative opinion of fever is very common—as if it were "the problem"

---

43   We can also have the students reflect on what the term "expression, to express" actually means.

rather than part of the solution. But according to a mainstream college biology textbook (Marieb & Hoehn 2012, p. 989): "Fever, by increasing the metabolic rate, helps speed healing and also appears to inhibit bacterial growth." Because the students have not yet learned about the immune system—they will in a few days—it is too soon to go into more detail. However, at a time when the anxiety around viruses is so great, it may make sense to mention at this point that "a higher temperature intensifies the effects of interferons" (Marieb & Hoehn, 2012, p. 798), which are protein substances produced by cells in response to the entry of a virus and that have the ability to inhibit viral growth. There would be more to say here regarding the relationship between temperature and immune system responses, but this small nibble is a way to whet the students' appetites for what is just around the corner (next week) in their very own 10th grade biology block!

So what do we have to thank for this amazing warmth organization that is so finely tuned to our bodily and consciousness needs? There are many factors, of course—it's an organism!—but a very central role is played by the heart. As students in 9th grade chemistry learned, warmth in the body is created through oxidation processes. When asked where in the body they think the most warmth is typically created, most students—based on their own experience of running and other forms of limb-based exercise—will think of the skeletal muscles. We can then present them with a chart such as the following, but leaving out the numerical data at first so they can try to rank them themselves (Kranich 2003).

Reflecting on the primary warmth creators before seeing the numbers above, some students may observe that if the skeletal muscles were actually

Warmth Creation in Major Human Organs

Organ	O_2 intake ml/100g/min.	Warmth creation ca/100g/min.
skeletal muscles	0.18	0.75
lungs	2	8.5
brain	3.2	15.4
digestive organs	3.6	17.4
kidneys	7	34.5
heart	10	45

the main source of warmth, then we would always get cold when we were not moving. Other perspectives may also come into the conversation, but lastly, when we fill in the numbers, they are usually amazed to learn what a central role in warmth creation the heart plays relative to other organs.

Looking back at what they have learned so far, they will agree that for the heart to access and process that much O_2, it must have a very dense capillary network. How dense is dense? Skeletal muscles have around 100 capillaries/mm.² To help them picture what that means, they are asked to draw a little square on a piece of paper with the dimensions 1mm x 1mm. Then they are asked to put 100 little dots in that tiny square, which represent the capillaries that would come out of the square if we were looking at a square-mm of skeletal muscle. It is hard to imagine such a density of blood vessels in such a small space (even though we learned how tiny they are earlier in the block). But if you have that many capillaries in a mm² of skeletal muscle, what will it be like in organs that produce much more warmth than those muscles do? Again, we provide them with a table such as follows, but leave out the numbers at first.

Capillary Density in Major Organs of the Human Body

Organ	Capillaries per mm2
Skeletal Muscles	100
Smooth Muscles	300–1000
Brain & Kidneys	2500–4000
Heart Muscle	8000 (Kranich)

Again, when revealed, the actual data is mind-boggling. 8000 capillaries coming out of the little mm² square we drew—unbelievable!

At this point, we can refer to what we learned earlier, that when the O_2 needs of an organ grow because of its increased activity over an extended period of time, that organ will build more capillaries to help meet those needs. "Getting in shape" has very much to do with just this. The capillary density will become greater in the heart and skeletal muscles of a person who exercises regularly, but they will thin out again if the person goes through an extended "couch potato" phase.

Part II. 10th Grade

Now, the amazing thing about the heart is that only about 20% of the O_2 it consumes is used for its basal metabolism, and only about 5% for the activity of beating (this can increase to 20% during physical exertion). This means that at least 60% of the O_2 consumed by the heart is turned into heat that is not needed by the heart, but is, instead, infused into the bloodstream to help warm the rest of the body (Lauboeck 2002). Because the blood contains so much water—which has the highest heat capacity of any natural substance—it can absorb a great deal of warmth[44] without getting much warmer itself and can carry this warmth to where it is needed in the body.

Coronary Circulation

After considering the huge role that the heart and circulation play in the human organism, it often makes sense—if time allows[45]—to look briefly at how the heart "takes care of itself." To start with, I like to bring an anterior and posterior view of the heart's coronary circulation, but without labels at first.

It provides a nice review and a good mental-picturing exercise for the students if they are asked to figure out where the major blood vessels (aorta, etc.) and chambers are located in preparation for labeling them. As we discussed

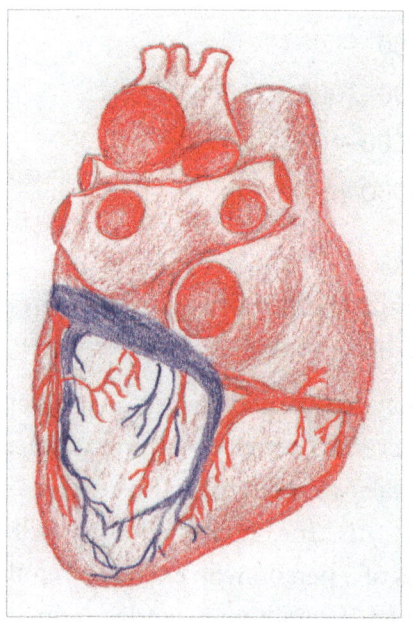

Fig. 2.21 Posterior view of the heart with coronary arteries and veins.

---
44 Human blood has 87% of the specific heat capacity of water.

earlier, the schematic two-dimensional drawings that illustrate blood flow sequence and valve locations do not accurately depict the three-dimensional spatial relations of the heart. So when they first see these two unlabeled images, many of them are baffled as to what-is-what, and must do some active inner translating of earlier images into what they now see.

Once they have achieved clarity at this level, various aspects of the heart-crowning (coronary) circulation can be considered. Thereafter, I usually go into some of the challenges and surgical corrective efforts that most of them will have heard about but have never really understood. For example, the so-called "hardening of the arteries" (arteriosclerosis, atherosclerosis) can be explained and illustrated.

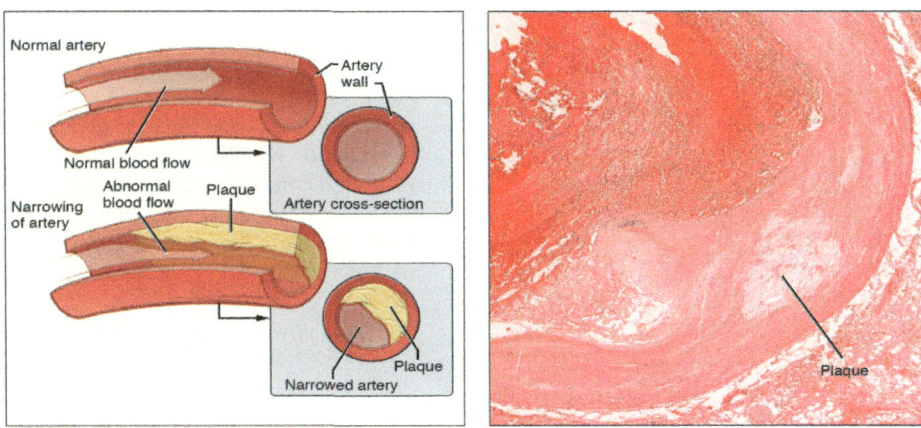

Fig. 2.22 Atherosclerosis: plaque buildup in an artery.
(Source: OpenStax College. Anatomy & Physiology, Connexions Web site. http://cnx.org)

---

45 As all Waldorf teachers know, blocks can vary in length from year to year, which influences how much time one can allocate to various aspects of a block. Some biology teachers may be surprised that so much time has already been allocated to the cardio-vascular system in this block. The perspective I follow in this regard was referred to earlier (in the Introduction) as "exemplary learning," or "less is more." The idea being that it is more fruitful for the students if we do not see our task as one of giving a tidy overview in the time available, but, to re-quote Martin Wagenschein, who says it so well: *Instead of evenly and superficially walking through the catalog of knowledge, step-by-step, we exert the right—or fulfill the duty—to really settle in somewhere, to dig in, to grow roots and take root. The particular aspect we delve into is not a stage in a process, but a mirror of the whole. Why? The relation the particular has to the whole is not that of a part, step, or preamble; it is a center of gravity. It may be only one, but it carries the whole in it.* In this spirit, I find it most effective to "settle in" to the heart and circulation and give them more time because they do, in my eyes, provide a "mirror of the whole" that we are trying to engage with in this block.

If this tendency goes too far, it can lead to a "coronary thrombosis" and result in a "myocardial infarction"[46] (heart attack). This can then be distinguished from what they have often heard of as a "stroke." Most students are also curious to find out more about procedures involving stents, balloon angioplasty, and coronary bypass grafts—things they have heard about but never understood.

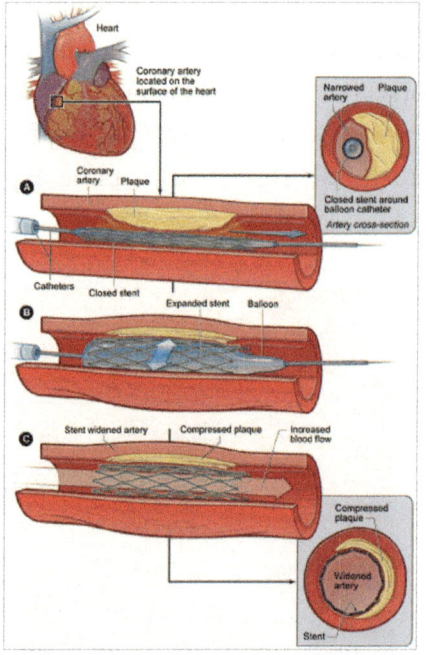

Fig. 2.23 Diagram of stent placement.
A. A catheter is inserted across the lesion.
B. A balloon is inflated, expanding the stent and compressing the plaque.
C. The catheter and deflated balloon have been removed. Before-and-after cross sections of the artery show the effects of the stent placement.
(Source: http://www.nhlbi.nih.gov/health/dci/Diseases/Angioplasty/)

Blood Pressure

We now take another look at what we have spoken about frequently in this block: blood pressure. We start with the difference between systolic and diastolic blood pressure, how a sphygmomanometer works, and the significance of Korotkoff sounds. At this juncture the teacher can offer to measure the blood pressure of several students and show the others how it is done. If interest is strong, a schedule can be set up to measure the blood pressure of everyone in the class and the results recorded in a table.

Based on our earlier discussions of the effect our inner life has on the heart and circulation, the students are interested to hear about "White Coat Hypertension," which refers to individuals who, when their blood pressure

---

[46] This term never fails to elicit laughter among the students; the reason why is probably obvious.

is taken in a clinical setting, grow nervous and show higher blood pressure readings than they normally would. This is related to the general phenomenon that when people are stressed their blood pressure spikes temporarily. This is due to an increase in heart rate and to vasoconstriction, which causes increased resistance to blood flow (www.mayoclinic.org/stress-and-high-blood-pressure).

Although we usually hear that normal blood pressure values are around 120/80 mmHg, in reality blood pressure levels are not uniform across different age groups. The following table shows how the average values for males and females change as we grow older.

Changing systolic/diastolic blood pressure values with age

Age (Years)	Minimum	Normal	Maximum
1 to 5	80/55	95/65	110/79
6 to 13	90/60	105/70	115/80
14 to 19	105/73	117/77	120/81
20 to 24	108/75	120/79	132/83
25 to 29	109/76	121/80	133/84
30 to 34	110/77	122/81	134/85
35 to 39	111/78	123/82	135/86
40 to 44	112/79	125/83	137/87
45 to 49	115/80	127/84	139/88
50 to 54	116/81	129/85	142/89
55 to 59	118/82	131/86	144/90
60 to 64	121/83	134/87	147/91

(Source: Medicalcharthelp.com)

Are we surprised? Although there are many factors that cause increases in blood pressure over time, one very common phenomenon is "hardening of the arteries" (arteriosclerosis, atherosclerosis). When arteries lose their flexibility and elasticity with age, they provide more resistance to blood flow and cause an increase in blood pressure. That humans lose flexibility in their bodies on many levels (and sometimes in their inner life) as they age is obvious to all. So it should not be a surprise when it happens in the arterial walls, as well. That the age-changes are much less evident in the diastolic pressure phase makes sense, since there is less diastolic flow to be resisted. The general trends are interesting to note here, although more comprehensive data would be necessary in order to draw more specific conclusions from this chart.[47]

Human Blood

Looking back over what we have considered in regard to the CVS so far, we can see that through the various phases and processes there has been one constant, one component that is always in play. That ever-present, always on-the-move organ is, of course, the blood. So it seems time that we take a closer look at it and see why it is such a central player in all that we have spoken about.

Since we have been following the blood in its moving dynamics up until now, it makes sense to see if an analysis down into its various components can help us better understand its various functions. With the help of a simple sketch, we describe how human blood, when spun in a centrifuge, divides into two main layers, a red one below that contains the denser-formed elements, the red blood cells (erythrocytes). At the top is the less dense, straw-colored plasma. The thin layer between, the so-called "buffy coat," is also made up of formed elements, the white blood cells (leukocytes) and platelets (thrombocytes).

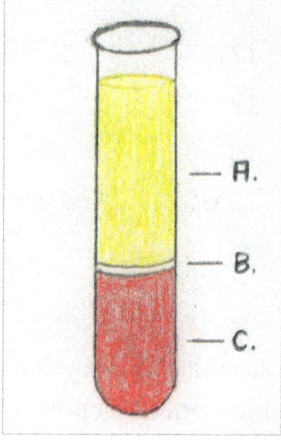

Fig. 2.24 Density layers that form in human blood after it has been spun in a centrifuge. **A.** 55% Plasma (= 90% water + protein). **B.** <1 % Buffy coat (leucocytes & platelets). **C.** 45% Erythrocytes.

---

47 For some classes it is a good idea to let them calculate the "pulse pressure" for each row (which is determined by subtracting the diastolic pressure from the systolic pressure) to show how much the gap between the two increases with age.

The Cardiovascular System

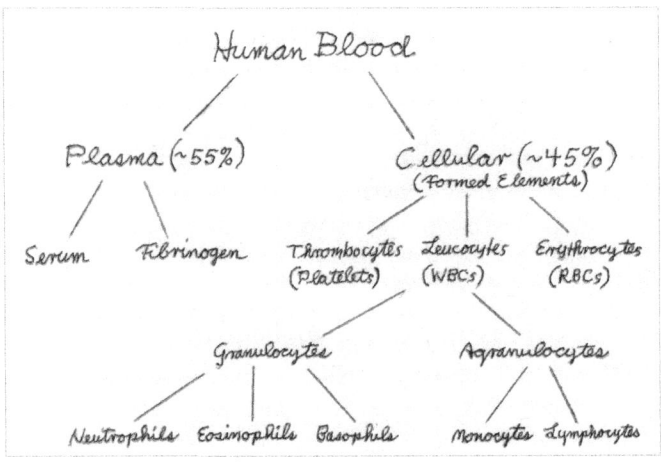

Fig. 2.25 Diagram of the primary components of human blood.

From this first analysis, a further schematic breakdown can provide a basic framework for further discussion.

As always, when entering into an area with countless details pertaining to it, the teacher has to continually strive to keep the forest in view (which is the larger pedagogical intent in covering this material) so as not to get lost in the trees that spring up in all directions. There is no "right" formula, of course, since the amount of detail that serves the students without overwhelming them varies from class to class.

Before going into the specific components found in the schema above, I have found it best to keep "the whole" alive by first describing the central functions of the blood in the human organism. To start with, it is important to note that because the blood permeates all organs and comes into contact with all their processes, it also reflects what is happening in them. It is for this reason that blood tests can be used to identify almost all pathological processes taking place in the organism (Rohen 2007, p.192).

One of the primary functions of the blood is often referred to as *transport and distribution*. In order to fulfill this task, the blood must be able to move, to flow, to circulate. To do so requires that the blood—in contrast to what we typically think of as organ systems—give up its specific form and location so that it can visit, permeate, and then move through the organs that are distributed throughout the body. Although the blood does possess formed (cellular) elements—see the diagram above—they are carried by the fluidity of the blood plasma throughout the body. In its transporting function the blood brings—as we know—O_2 from the lungs to the bodily tissues and CO_2 from the tissues to the lungs for exhalation. It also transports nutrients from the gastrointestinal tract to the bodily organs, as well as hormones from the endocrine glands to targeted locations. It transports metabolic wastes to certain organs for elimination, and very importantly, as we have heard, it transports heat throughout the body.

In this context, it is important to help the students realize that—to paraphrase Wordsworth—we can "distinguish" between these functions, but should "not divide" them. That is to say, these functions do not run autonomously and parallel to each other, but are interwoven. To name a function is to emphasize an aspect, not to isolate it. For when you have life, you have intersecting factors and functions, not clearly separate causal tracks.

A second central function of the blood is *regulation*. The blood keeps physiological conditions in the organism in balance (homeostasis). This includes heat distribution, insuring a healthy fluid-volume of blood, and maintaining the slightly alkaline pH levels of the blood through so-called "buffers."[48]

The third large functional category—the so-called *protective function*—involves helping preserve the organism's integrity by preventing blood loss through coagulation, on the one hand, and by warding off influences foreign to the body that threatens its biological integrity, on the other.[49]

Such an overview provides a context for highlighting the blood components that appear on the overview diagram. (You have the basic forest, and now you can look more closely at some of the trees.)

Beginning with the all-important, life-enabling red blood cells (erythrocytes, RBCs), the students are amazed to learn that we have around 25 trillion in our bloodstream. The fact that there are approximately 5 million contained in one cubic mm (mm3, microliter),[50] translates into some 270 million RBCs

---

[48] Many of the chemical reactions that occur in the body, in particular those involving proteins, are *pH*-dependent. Thus, human blood contains a buffer of carbonic acid (H_2CO_3) and bicarbonate anion (HCO_3^-), which serves to maintain blood pH between 7.35 and 7.45.

[49] This last area involves the immune system and is immense. We will touch upon some of its activities in regard to the blood here, but will address it again later as its own topic.

[50] Such figures tend to vary according to gender: Women typically have an RBC count of 4.3–5.2 RBCs/mm³ of blood, whereas men usually range from 5.1–5.8 RBCs/mm³.

in a drop of blood! What's more, they are not only small (Ø7.5 micrometers), but with over 2 million expiring and being replaced every second, they are in a constant process of dying and becoming.

This dynamic allows them to adjust to changing environmental conditions very rapidly. We ask the students if anyone has ever climbed a mountain shortly after coming up from a lower elevation. Some, often many, will report the experience of being rapidly out of breath, but that after a week or so they find themselves much better acclimated. What changed in the meantime? The number of RBCs in their bloodstream! At high altitude there is less air pressure and thus each RBC can bind less O_2 than it could at lower elevations. Less O_2 per RBC means less O_2 intake and leads rapidly to shortage of breath. But the body begins right away to adjust to this circumstance and to produce RBCs at a higher rate than the number that are dying. In about a week, one is normally about 80% acclimated when, for example, coming from the Midwest to the Colorado Rockies. Because of the increase in RBC count that develops at high altitude, many athletes will train in such areas and then return to lower altitudes when it comes time to compete. They now have the advantage of being able to bind more O_2 with each breath, thanks to the higher RBC count they acquired in the high country (which will decline again if they stay at lower elevations).

Students will have heard of illegal doping in the context of competitive sports. One form of this is to have some of an athlete's RBCs drawn off, stored, and then re-injected a few days before the competition takes place. (Question for the class: Why does this help, aren't they just getting back what they gave away with no net gain?)

RBCs also follow a very unusual developmental path. During their formation in the red marrow of our bones, the erythrocytes-to-be begin to synthesize a substance known as hemoglobin.[51] As they progress further, something unheard of in normal cell development takes place: when almost all of the hemoglobin has been produced and it begins to turn pinkish, the cell ejects most of its organelles, as well as the cell nucleus.[52]

---

51 Hemoglobin contains iron, which has the unique property of combining reversibly with oxygen, which makes it an ideal medium for transporting oxygen, and to a lesser degree carbon dioxide.
52 Since the students have not yet studied cytology, short explanations to "organelles" and "cell nucleus" may be necessary. At the same time, it can be pointed out with great excitement that next year they will learn lots and lots about cell organelles and the roles they play in the life of cells, which, in turn, have a role to play in the life of the organism of which they are a part.

Part II. 10th Grade

This has huge consequences. The anucleate RBCs are unable to grow, divide, or synthesize new components to replace damaged ones. As a result, when they grow old they lose their flexibility and become more and more fragile, which leads to death in about 120 days. (After death they are broken down and their iron is "recycled.") Now that we know the life span of an RBC, we can give the mathletes in the class a straightforward computational task. If, as we heard earlier, we have about 25 trillion RBCs in our bloodstream and they live an average of 120 days—meaning they all have to "die" and be replaced in that period of time—then how many must die and how many "birthed" every second of every day and night our entire life long?[53] Around 2.4 million RBCs must die and be replaced every second! Wow, that is quite amazing, the students agree. If we have been in class for an hour at that point, we can let them figure our how many RBCs have died and been replaced since the lesson began. Result: about 8 billion 640 million—in just one hour! What an unbelievably dynamic world lies beneath our skin! But, hopefully, the students will feel that this is in sync with all we have already learned regarding the dynamic nature of the CVS in this block.

Is there method in this madness? To what end is this sacrifice of life-sustaining cell components? It creates the possibility for the RBCs to serve the rest of the organism as the transport vessel for O_2. For starters, when the cell nucleus is released, the cell collapses inward and assumes a biconcave, miniature donut-like shape that has a huge surface-area relative to volume. With no point far from the surface, it becomes ideal for gas exchange. As it turns out, the total surface area of the RBCs in an average adult is 3800 m 2, which is 2000 times larger than the surface area of our skin! That's a lot of surface to breathe with! Moreover, because they can generate energy-rich ATP[54] anaerobically, they do not consume any of the O_2 they carry while transporting it. And, the biconcave (donut-like) shape enables RBCs to deform without rupturing when squeezing through narrow capillaries.

Turning now to the second major group of formed elements, the white blood cells (leucocytes, WBCs), the students can easily go pale when deluged by a series of—for them—opaque terms (neutrophils, etc.) and the many details that can accompany them. In fact, even for experts leucocytes are difficult to

---

53 25 tril. / 120 d. / 24 hrs. / 60 min. / 60 sec = ~2.4 mil./sec.
54 Another topic for 11th grade.

The Cardiovascular System

classify because of their extremely complex—and at the same time—flexible, structures. Most also produce chemicals of many kinds. Moreover, immune responses involve a good deal of cell-to-cell communication.[55]

To avoid overwhelming the students, I have found it beneficial to emphasize several primary functions and ways of being that we find among this group of cells, and then use the names as a way of making clear that we are looking at a kind of "division of labor" here.

In that spirit, a good way to start off is by noting that the WBCs come from the same red bone marrow[56] as the RBCs, but with several significant differences. On the one hand, they contain no iron or hemoglobin; on the other, they retain their nucleus, organelles and cytoplasm. As we will see, their life spans and abundance vary, but taken together, there is, on average, only about one WBC for every 700 RBCs. That said, they are present everywhere and always ready to go into action. Despite their differences, they have one function in common: they ward off foreign substances and pathogens that endanger the integrity of the organism.[57]

In contrast to the RBCs, WBCs do not remain within the blood vessels, but slip out of the capillaries (diapedesis) after using the blood circulation to transport them to areas of the body where they are needed. Once out of the bloodstream, they use amoeboid motion (the teacher can try to graphically describe and demonstrate, what this is like) to follow the chemical trail of damaged cells, or of other WBCs. Arriving at areas of endangered tissue in this way, they gather there in large numbers to destroy foreign substances. Whenever they are mobilized for action in this way, the body speeds up their production and within a few hours twice the normal number may appear in the blood.

Of the WBCs found in our diagram, the primary players active here are the granulocytes (neutrophils, basophils and esosinophils). They are called

---

55 Rather than lay all this out in detail for the 10th graders, the goal is much more to convey a sense for the interwoven, coordinated working in this realm that is amazingly intricate and marvelously effective. This impression can grow stronger when we address the specific (adaptive) immune system a few days later.
56 From the same hematopoietic stem cells
57 This topic will be taken up in more detail when we cover the immune system, as such, later in the block. For now, our focus will be on the non-specific (innate) defenses only.

granulocytes because of the small granules within their cytoplasm that contain little packets with very specific chemical contents. During a process known as degranulation, which is triggered by the immune system, they move out of the cell and break apart, releasing their content into the interstitial space. Depending on the type of granule, they then perform a range of immune functions.

The neutrophils make up 50–70% of the WBC population. Attracted to inflammation sites, they are usually the first to arrive and begin immediately to envelop foreign matter and digest it. The students usually enjoy some kind of exaggerated description of what this process would be like. For instance, if their classroom were part of an organism and the students themselves were invading bacteria who had snuck in from a "neighboring" high school. (Please note, the sensitive reader may wish to skip the rest of this paragraph.) Soon the neutrophils would pick up on this and move amoeba-like from hallways all over the school building into the classroom, where—the students now try to picture this—the students themselves would be enveloped in amoeba-fashion and digested. Known as "bacterial slayers," the neutrophils would increase in number explosively and the remains of the 10th graders would soon appear as pus seeping out the classroom windows. A crass—and far too simplified—description of this sort should be done playfully, of course, and would also need to be varied depending on the class in front of one, but something along these lines helps the students get a better sense for the nature of phagocytosis and what results therefrom. The word "pus" is a trigger word, of course, and must be framed to fit the group of students one is teaching. When you engage the students' feelings—even in a rather "over the top" manner such as this—what is being discussed stays with them more strongly. As Rudolf Steiner emphasized in his lectures *Education for Adolescents* (1996, CW 302):

> *It is our life of feelings…that is the actual vehicle for the enduring qualities of ideas and mental images that we can recall at a later stage. Our mental images change into stirrings of feeling, and it is these stirrings of feeling that we later perceive and that enable us then to remember. … Their memory will be greatly enhanced … if we spice our lessons with the possibility of allowing the children to experience corresponding emotions, if we make them smile or feel sad. … We need not be pedantic in this teaching, need not always connect feeling directly to the subject taught. We may refer to something else in order to*

stimulate feelings. The important thing is that the children's feelings are engendered during a lesson. This stirring of feelings aids memory (pp. 18–19).

Since Steiner made such statements, the connection of feelings to memory and learning has been explored extensively by psychologists and neurologists, who have confirmed that the engagement of feeling influences attention and memory significantly.[58] The amygdala in particular—a brain structure crucial for the formation of memories—comes into play when the feeling life is engaged (Restak 2006).

Esosinophils make up only 2–4% of all WBCs but play the important role of attacking parasitic worms that burrow into the intestinal or respiratory walls and are too large to be enclosed by phagocytes. Instead, the esosinophils gather round the "prey" and release enzymes onto the parasite's surfaces that digest it away.

Most students have had experiences with mosquito bites and—finding them itchy and the source of an unaesthetic red bump on the skin—tend to blame the mosquitoes. Now they learn that instead of blame, they should be thankful. But not to the mosquitoes, but rather to the basophils, our third main granulocyte. Although basophils make up only 1% of the WBCs, they play the significant role of producing histamines, an inflammatory chemical that dilates blood vessels and makes them more permeable to WBCs. This is important because mosquito saliva contains proteins that most people are allergic to. Through dilation of the blood vessels—which causes the unattractive red bump—the WBCs can spring into action much more effectively. However, the histamines also irritate nerve endings in the area, which leads to itching. And the students should be thankful for this too, because otherwise they would find themselves covered with little unattractive red bumps that they hadn't noticed as they occurred because they didn't itch. Because they did not know whence (and when) the unsightly bumps came, they would not think to apply their favorite organic mosquito repellent in a timely fashion to prevent further red-bumping.

---

58 See for example Tying et al. in "The Influences of Emotion on Learning and Memory." They note, for example, how "numerous studies have reported that human cognitive processes are affected by emotions, including attention, learning and memory." They conclude that "this knowledge may be useful for the design of effective educational curricula to provide a conducive learning environment" (2017).

Monocytes are agranular and make up 7–8% of the WBCs. They travel to inflammation-related stimuli and begin producing phagocytic daughter cells—that are crucial to defending against viruses and chronic infection—in great numbers. If they encounter something too big for one of them to engulf and digest, a bunch of them joining together to form a "phagocytic giant cell," which gets the job done!

The final—and eminently important—group of WBCs to be mentioned is the *lymphocytes*. The students will have heard of the immune system of course, but probably not that the immune system has two "branches," so to speak, that need to distinguished. The first, to which the lymphocytes do not belong, is the *innate (non-specific) immune system* that we have been referring to up until now. The non-specific system is always ready and able to react when pathogens enter the body. The *adaptive (specific) immune system* (to which the lymphocytes belong) works differently. It protects us from a wide range of infectious agents (antigens), but requires a "biological learning process" for it to be effective. Because it is intimately interwoven with the lymphatic system and is involved in many "hot topics" such as vaccinations, AIDS, and pandemics, I have found it more effective to cover this group of WBCs a bit later and more extensively under the larger rubric of "the immune system."

Rounding off the group of formed elements in human blood are the *platelets*. Platelets are not actual cells but cell fragments.[59] They, too, are involved in maintaining the integrity of the organism—albeit in a very different way than the WBCs. They play a central role in a process called "hemostasis" (prevention of blood loss)—without which we would quickly lose our entire blood volume through even the smallest wound.

As soon as a blood vessel wall has been damaged, hemostasis begins with a 30-minute period of vasoconstriction. In large vessels this reduces blood loss through a decrease in flow volume; in capillaries, it stops blood loss altogether.

In a second phase, that begins within 30 seconds of the cell wall injury, platelets attach to exposed collagen fibers in the damaged vessel wall. Chemicals released when the platelets break open, as well as chemicals in the plasma,

---

59 They are often referred to as "thrombocytes," although some disagree with this moniker since platelets are not true cells. During blood formation in the red bone marrow, the largest cells there, the megakaryocytes, each splinter off some 2000–3000 fragments—the platelets—that are filled with substances essential for blood clotting.

stimulate further vasoconstriction and the aggregation of platelets, which serve to plug up the hole in the vessel wall.

The concluding coagulation phase also begins within the first 30 seconds and involves a cascading series of reactions (like falling dominoes) of multiple clotting factors (coagulation factors). The most important result of this process is the chemical conversion of the substance fibrinogen[60] from its dissolved state in the blood plasma to a network of insoluble fibrin fibers that trap the WBCs and RBCs, thereby sealing the hole in the injured blood vessel until it can be permanently repaired (Lazaroff 2004, Marieb & Hoehn 2012, Totoro & Derrickson 2013).

This brings us back full circle to the fluid element of the blood—the plasma—which links everything together that we have spoken of so far. As our chart and test-tube drawing indicate, the plasma makes up 55% of the blood's volume. Consisting of almost 92% water, it is such an excellent solvent that it is able to transport a diverse range of dissolved substances (solutes) throughout the body to the locations where they are needed. The majority of those solutes are plasma proteins (albumins, globulins, fibrinogen). The remaining 1.5% is made up of other solutes such as electrolytes, nutrients, hormones, respiratory gases and waste products. All of this—and the formed elements—is carried by our great internal "Mississippi" and its watershed—which also includes a whitewater segment—throughout our inner landscape and makes the interwoven web of life in the human body possible!

Since most classes will have learned about *blood groups* and the *Rh factor* in grade school, a brief reference to them is usually sufficient before moving on. If they are not familiar to the students, I have found an historical approach to these topics to be effective, referring back to how dangerous and frequently fatal blood transfusions were prior to the work of Karl Landsteiner. It is interesting for the students to reflect on the kind of experiments Landsteiner did in order to identify the blood groups and determine how they relate to each other.

It is now time to proceed to one of the hot topics of our times: the human immune system, and to look at some of the major challenges it has faced over the past half century.

---

60 The pivotal coagulation factor for this conversion is the enzyme thrombin, which itself arises out of the cascading reactions.

The Immune System

The Nonspecific (Innate) Immune System

There are two levels of the human immune system that we need to distinguish. The nonspecific (innate) and the specific (adaptive). Although we have already covered most of the internal aspects of the nonspecific system in our discussion of the WBCs, there is also an external factor to be considered. Often referred to as the "first line of defense" provided by "surface barriers," it involves the skin and mucous membranes, along with the secretions that the membranes produce. Bacteria rarely penetrate an intact skin surface. In addition, the viscous mucus produced by the membranes that line our body cavities traps many microbes and foreign substances. Without getting lost in too much detail, several fluids that are part of this first barrier should be mentioned, such as tears, saliva and urine flow. Sebaceous (oil) glands in the skin also secrete a substance called sebum that forms a protective film over the surface of the skin that inhibits the growth of certain pathogenic bacteria and fungi. Even perspiration helps flush microbes from the skin's surface.

It is particularly important for the students to realize that the skin is actually a complex ecosystem of diverse habitats that is colonized by a rich collection of microorganisms. Beyond the 100 billion bacteria (of which there are approximately 1000 different species) that inhabit our skin, we also host fungi, viruses, and even mites. Symbiotic microorganisms are housed in a wide range of skin niches, where they help prevent transient pathogens from colonizing the skin surface by consuming their nutrients, secreting chemicals against them, and by stimulating the skin's immune system. The delicate balance between host and microorganisms is an important topic of discussion in a culture where the use of "antibacterial soap" is often excessive[61] (Sender, Fuchs & Milo 2016; Grice, Kong, Conlan, et al. 2009; Grice & Segre 2011; Akst 2014; Stromberg 2014).

The "second line of defense" is needed when pathogens penetrate the surface barriers. It is internal and has both nonspecific and specific components. Central to the internal nonspecific system is the phagocytic activity of the WBCs

---

[61] Prior to 2016, one of the primary components in antibacterial soaps was triclosan, which was banned in 2016 by the FDA because of its negative effects at several different levels.

(neutrophils, monocytes) we discussed earlier, and the activity of a group of lymphocytes in the blood known as "natural killer (NK) cells" that attack body cells that show unusual or abnormal membrane proteins—for example virus-infected body cells and certain cancer cells.

The interior nonspecific defense system also includes:

- Antimicrobial substances, such as interferons (IFNs) that discourage microbial growth.
- Inflammation—redness, swelling, heat, pain (a nonspecific response that we referred to earlier in regard to histamines with our mosquito example).
- Fever, too, plays an important role. Elevated body temperature not only intensifies the effects of interferons and hinders the growth of some microbes, but it also increases the speed of reactions that aid body repair. Because fever is so often misunderstood and suppressed in the United States, I quote below two college-level biology textbooks about its importance.

Although death results if core temperature rises above 112 to 114° F, up to a point fever is beneficial. For example, a higher temperature intensifies the effects of interferons[62] and the phagocytic activities of macrophages[63] while hindering replication of some pathogens. Because fever increases heart rate, infection-fighting white blood cells are delivered to sites of infection more rapidly. In addition, antibody production and T cell proliferation increase. Moreover, heat speeds up the rate of chemical reactions, which may help body cells repair themselves more quickly. (Tortoro & Derrickson 2013, p. 1012)

In order to multiply, bacteria require large amounts of iron and zinc, but during a fever the liver and spleen sequester these nutrients, making them less available. Fever also increases the metabolic rate of tissue cells in general, speeding up repair processes (p. 798). As body temperature

---

62 A protein released by animal cells, usually in response to the entry of a virus, which has the property of inhibiting virus replication.
63 A large phagocytic cell that is found both in stationary form in the tissues and as a mobile WBC that travels to infected sites.

rises, enzymatic catalysis is accelerated: With each rise of 1°C, the rate of chemical reactions increases about 10% (p. 985). Fever, by increasing the metabolic rate, helps speed healing, and it also appears to inhibit bacterial growth. (Marieb & Hoehn 2012, p. 989)

The Specific (Adaptive) Immune System

After encountering the sea of details we have just covered, the students need to breathe out a bit. A refreshing way to do this, and to introduce them to the specific immune system at the same time, is to look at several historical events that paved the way for our modern understanding of biological immunity and how it comes about. I am referring to the well-known "experiments" performed by Eduard Jenner and Louis Pasteur involving vaccination and vaccines.

These stories can be described quite graphically (which I do not do here). Jenner, for example, decided to test the local folklore that milkmaids who suffered the mild disease of cowpox (an occupational hazard!) never contracted small pox, one of the worst killers of that historical period (which had a fatality rate of 20 to 60 percent and usually left survivors disfigured and blind). He tested his theory by inoculating an eight-year-old orphan, James Phipps, with the content of cow pox lesions from the hands of a milkmaid. The boy developed a fever, but eight days later, when Jenner inoculated him with substance from a fresh small pox lesion, no disease developed. Pasteur, almost a century later, inoculated 25 cows with attenuated anthrax bacteria, while a control group of 25 was left unvaccinated. Thirty days later both groups were injected with a culture of live anthrax bacteria. All the animals in the vaccinated group survived, all the animals in the non-vaccinated group died (Van de Graaff & Fox 1998). The stories are both fascinating and repelling (ethically) and help the students never forget where the term vaccination (*vacca* = cow) comes from.

The oft-mentioned danger of losing sight of the forest for the trees is greater than ever when we move into the specific (adaptive) immune system. The details and accompanying terminology become overwhelming for 10th grade students if the teacher does not find a way to keep "the essential" in mind, while supplying just enough specifics to let the overriding whole shine through.

What does the work of Jenner and Pasteur show us? This we can discuss with the students. Evidently both examples start with the first-time encounter with a pathogenic substance. It was cowpox antigens in the case of the milkmaids. That first meeting has consequences: they get cowpox, which is a minor illness for humans. In the future, when they come into contact with cowpox they will not get it again because their bodies have a) "learned" how to recognize it, b) learned how to deal with it, and c) will "remember" these things into the future. What is important about this example is that once their body's immune system learned to deal with cowpox, it became able—because cowpox and smallpox are both members of the pox family—to deal with the more ominous sibling, smallpox, as well.

It is a big help to the students if we turn what they have just heard into a somewhat "silly" anecdote, so they can better picture and remember the immunity-development process. For example, if we want to follow up on the cowpox-smallpox immunity narrative, we can tell them about two siblings, Bob and Bill, who were less than forthright salespersons. Let's say little bro Bob was a charmer, who got you to fall for his sales pitch and to pay way more than you should have for your new tennis racket. Bummer. But you learn from this mistake and when you go to buy a new car a year later, from (as it turns out) brother Bill, you recognize his phony sales pitch immediately and know how to counter it with skepticism and by making a lowball counteroffer. No phony salesperson will ever be a problem for you again because, after your painful first experience buying a tennis racket from Bob

a. you now recognize a phony sales pitch,
b. you have figured out how to deal with it and move forward to your benefit, and
c. you will remember long into the future what such pitches are like and what you will need to do to counter them.

Unfortunately, this will not protect you if you run into a sweet-talking "charmer" at the local pub, but you will go through a similar learning process, which will make you immune to this variation of the "manipulator" theme in the future, too!

Far-fetched? Yes,[64] but if the students get into it and feel it (finding it funny, corny, clichéd, absurd), then they are much more likely to remember it later.

Part II. 10th Grade

The important question now becomes: What is the primary difference between the kind of immune system we learned about when we studied the workings of the WBCs in the blood (the "nonspecific," innate system) and what we have just been talking about and are calling "specific" (adaptive) immunity? One possible way to characterize their relationship is to compare the nonspecific system with street cleaners, who move down the alley behind your house and remove every bit of garbage they find along the way. This is similar to the phagocytes, that go after any foreign substance they meet, then envelop it and digest it. "Garbage is garbage." Specific immunity, by contrast, is much more like a health department specialist, who has the specific task of ridding your alley of rats, for which they use precisely-determined poisons and traps that they have developed based on past experience, and that should (!) kill rats and only rats. These specialists: 1) recognize immediately the tell-tale signs of rat presence (rat poop around the garbage cans); 2) can recruit and train the co-workers they need to meet the issue lickety-split; and 3) jump into action immediately to maintain our beloved rat-free alleyways as if no rat had ever been there. (Whatever alleyway garbage still remains is not their problem.)

Of course, the actual process in the immune system is much more complex and the factors required much more interconnected than our simple examples above. The critical pedagogical question at this juncture is: How much detail do we need to guide the class through this highly complex realm without awakening an immune reaction in the students that could cause them to turn away from such topics forever—assuming they had survived this first meeting with only a headache and a bad case of hypersomnia (extreme sleepiness)?

There are many ways to attempt this antigen-rich journey, of which the following is just one. First of all, we can try to clarify the basic task of the specific

---

64 How can recognizing and dealing with a phony salesperson be related at any level to what immunologists are referring to when they speak of a T cell receptor (paratope) that fits (binds) perfectly with the corresponding part (epitope) of one particular antigen (analogous to a lock and key)? The comparison of such a process with what goes on in human learning appears less absurd (more "fitting") when one considers how all learning involves making judgments. A judgment is made when a concept is brought forth that "fits" the percept one is engaged with. If the concept "fits" it sheds light on that percept and enables one to understand it and behave accordingly (fittingly) when interacting with it. (For every percept we seek a "fitting" concept, and until we have found it our behavior in regard to it will not be fitting—which can cause us, on occasion, to throw a fit!)

immune system, which is to defend the body's integrity against large numbers of very specific invaders, each of which must be dealt with differently. It is therefore necessary that three capacities exist at this level of the organism:

1. The invaders (called antigens) must be noticed ASAP and identified as foreign and invasive.

2. The way to deal with each specific antigen must then be determined and multiple new cells—called "effector cells"—produced that can do this (because there will be many of these specific invaders coming at once).

3. But since this is usually not a one-time-occurrence, there must be cells that do not get involved in the destruction process, but quasi "stand back" and keep an eye out for the reappearance of that specific antigen in the future. They remember what it looks like and are thus called "memory cells." If they spot this antigen again, they quickly react by reproducing and differentiating into more effector cells and more memory cells, in order to deal with it quickly and thoroughly.

These three steps describe very roughly the ongoing task of the specific (adaptive) immune system.

Next, we must learn a bit about the means that the human organism[65] has developed to fulfill this task. We need to return to the WBCs in order to get a somewhat (but only somewhat) closer look into this complexity. Earlier we discussed several kinds of WBCs (neutrophils, etc.), but left out the lymphocytes —even though they appeared on our diagram—saying we would return to them later. Well, later is now (now is the new later). The students can be introduced to the extremely complex specific immune system in a very simple fashion through the two main types of lymphocytes: the B-lymphocytes (B cells) and T-lymphocytes (T cells). Both originate in the red bone marrow. The B cells complete their development in the bone marrow (hence the B), while the T cells migrate to the thymus gland and mature there (hence the T). These lymphocytes and their further differentiations are at the center of the three tasks described above.

---

65 Since it has a much more complex immune system than any other vertebrate or invertebrate.

B cells are active in all three tasks, but in a different way than the T cells. When a B cell meets an antigen and recognizes its invasive character, it produces great numbers of "antibodies" (immunoglobulins)—in a process known as clonal selection[66]—which are secreted into the blood and lymph and serve as "effector cells" to destroy extracellular antigens located in the bodily fluids (humoral immunity; antibody-modulated immunity). Because an antibody can only bind an antigen that has a complementary shape, the metaphor of lock and key is commonly used here. Just as a key fits only one lock, an antibody binds only its complementary antigen, and once that happens, the antigen can be inactivated in various ways.

Parallel with this process, long-living B "memory cells" are produced that continue to circulate for years keeping a watch out for the return of that specific antigen.

In a process called cell-mediated immunity, T cells undertake the same three tasks, but are focused on antigens found within cells (bacteria, viruses, fungi; some cancer cells; and tissue transplants). Rather than creating antibodies, the T cells themselves proliferate and differentiate (clonal selection) into specialized "effector" and "memory cells." As a result, there are millions of different T cells in the body with the unique ability to recognize (memory cells) or to deal with (effector cells) one specific antigen.

Depending on its location, an antigen can incite both types of immune responses. This is because when a new antigen enters the body there are many at once. Some will move into the body cells (provoking T cells), while others will be present in the extracellular fluid (provoking B cells). As a result, both types of immune response often work together to get rid of one particular antigen (effector cells) and both remain alert into the future (memory cells) in order to rapidly deal with an antigen should it reappear.

It is often surprising and interesting for the students to learn from all this that childhood means going to school on more than one level. Born without a developed immune system, the baby is protected from total exposure to outside influences by the mother's milk, which contains antibodies that protect it, as

---

[66] Swollen lymph nodes in the neck, for example, are usually caused by this sudden production of lymphocytes.

well as through antibodies transferred from the mother to the baby through the placenta before birth. After that protection ceases, the child has to develop its own immune system by interacting with the outer world. Their immune system is constantly learning through all the encounters the child has with its environment. Fevers play an important role in this by mobilizing both the humoral and cellular defenses. Through their meetings with antigens from without, the lymphocytes are activated to develop the exact counter gestures (lock and key) to all the substances they meet. This process is most intense in the childhood and early teen years, when the thymus gland is at its largest and most engaged in the task of educating lymphocytes. After puberty it begins to atrophy gradually, until by old age it is essentially gone.

If you assume that a large proportion of the estimated 10 billion specific antibodies of the adult humoral system are developed during the first 15 years, then it gives you a sense for how active the child's immune system is in these years. It is continuously asking as it meets new substances: Does this belong to me or does it not belong, can I take it in and make it part of myself, or do I need to reject it and protect myself from it? This ongoing life process creates the physiological foundation for the body's own identity, for what is often referred to as the "biological self" (Schoorel 2004).

Interestingly, the young person goes through a similar process biographically in the way they gradually learn to shape their life and behavior consciously and independently, as they become able to distinguish between "the good, the bad, and the ugly."

Before moving to the devastating example of what happens when our specific immune system is not functioning adequately (HIV-AIDS), I usually try to ground the kids a bit by rounding off the current considerations with a few more concrete issues that the students have heard of or experienced themselves, such as allergic reactions, auto-immune disease, mononucleosis (mono: "the kissing disease"), swollen glands, etc.

This can be followed by a discussion of the role played by vaccinations (passive and active) in our time. This is a hot topic in the United States, and I find it important to introduce the students to several different perspectives on this issue so that they can transcend the kinds of "nothing but" judgments that are either "for" or "against," with no nuanced options in between.

As always, we try to consider the relationship between our inner life and what takes place on a biological level before concluding such a topic. Research in the field of psychoneuroimmunology has confirmed what people have long observed: that your moods, feelings and thoughts influence your health. For example, cortisol, a hormone secreted by the adrenal cortex in connection with a stress response, inhibits immune system activity. It has been found that the loss of a close acquaintance or relative can result in a measurable weakness in the immune system. Loneliness, depression and fear are frequently connected with a decrease in certain lymphocytes (T helper cells). On the other hand, psychosomatic research has shown that the immune system is more effective, the more idealistic, optimistic, and courageous an individual is (Albonico & Lehmann 1993, Lancet 1987, Tortoro & Derickson 2013, Wolff 1993).

HIV/AIDS

One effective way to round off the study of the human immune system is to give an example of what happens if only one element in the web of interacting factors is hindered from playing its part. Tragically, this has been demonstrated world-wide on an enormous scale since the advent of HIV/AIDS in the early 1980s.

Because the students have all heard of HIV/AIDS and have some level of understanding of what it involves, they are usually interested in hearing a bit about its history, about how it suddenly appeared—seemingly out of nowhere[67]—in 1981, when a small number of hitherto healthy young men in California were diagnosed with pneumonia caused by a parasite (*Pneumocytes carinii*) that normally occurs in rodents. Over that summer more and more rare infections were reported in males of that age group (30–39). Such illnesses are called opportunistic because they take advantage of the opportunity to infect a human being—something a healthy human immune system would not allow them to do. In other words, cases were popping up more and more where people were becoming susceptible to diseases normally found only in animals.[68] The story unfolds with many interesting and disturbing details if the teacher chooses

---

67 Although it was later learned how an unnamed disease with similar characteristics had been reported in Africa in the mid 1970s, further research traced it back to the 1940s.
68 As AIDS spread further in the United States, it was found, for example, that when a person with AIDS developed tuberculosis, it was—almost without exception—bird tuberculosis (*Tuberculosis avi*) (Bos 1989).

to pursue these further. I will not describe that unfolding here, but simply note that from the first five cases reported early in 1981, by the turn of the century 22 million people had died of AIDS worldwide, and 35–40 million more were HIV positive (contained antibodies to HIV in their blood).

When the Human Immunodeficiency Virus (HIV) was finally isolated and identified, it became clear that the virus infects and kills a specific type of lymphocyte, the "T helper cells" (CD4 cells). These cells play a central role in the human immune system. Rather than taking on invaders themselves, T helper cells function more like team coordinators. When activated by the presence of an antigen, they spring into action by secreting a variety of small protein hormones known as cytokines, which play a large role in the activation of other T cells and B cells. For example, one cytokine produced by T helper cells, interlukin-2, is needed for virtually all immune responses. It is not only the central trigger for T cell proliferation, but it also both activates and causes a huge increase in the number of B cells and NK cells.

It would be easy to get lost in the details here, but one important takeaway is helping the students see how the immune system is a complex intertwining of many factors at many levels and that certain key players are pivotal for all of the others—just as every organization or team has key members who are pivotal in determining if they succeed or fail. When a key player like T helper cells in the human immune system is weakened, then the amazing web of coordinated immune reactions begins to grow weaker and the human being gradually loses the capacity to retain their own unique "biological self."

The normal T helper cell count is between 500 and 1500 cells/ml., but when an individual is infected by HIV that count begins to drop. When the cell count falls below 200/ml the person is said—by definition—to have AIDS. To make the significance of a decreasing number of T helper cells more concrete for the students, here is a list the symptoms that commonly accompany decreasing numbers of T helper cells (CD4 cells).[The normal CD4 count is between 500 and 1500 cells/ml.]

- CD4 count < 350: skin ulcers; tuberculosis; oral or vaginal thrush (yeast infection); cancer of lymph glands; skin cancer (Kaposi's sarcoma)
- CD4 count < 200: pneumonia; skin lesions; yeast infection of esophagus

Part II. 10th Grade

- CD4 count < 100: meningitis; yeast infection of brain lining; AIDS dementia; encephalitis: brain infected by a parasite often found in cat feces; weight loss (wasting syndrome); extreme diarrhea
- CD4 count < 50: blood infections; viral infections of almost any organ system; blindness

In order to bring home the actual tragedy of this decline in the immune system, I often read the students real-life descriptions of a person with full blown (third-stage) AIDS. One example would be the following by Dr. Arie Bos:

I can well remember the first time I stood face to face with an AIDS patient ... His "buddy"...had phoned me asking whether I was willing to be his doctor, and if I could come over immediately. I found a completely emaciated man, confined to bed and continuously scratching himself. His sunken cheeks and hollow temples gave an unusually sharp prominence to his cheekbones. The deep eye sockets contained a pair of vulnerable, almost fearful eyes. His pale skin was covered with festering sores, a result of incessant scratching. His hair was thin and dull, his mouth and tongue covered by a white coating. ... In a monotonous voice, he kept on repeating the same short sentence: 'isn't it awful?' It did not sound as if he meant anything in particular with this question. (1989, p. 53)

Most striking about a person with full blown AIDS are the pronounced cheek bones and hollow eyes. The large eyes appear quite vulnerable. The skin in some cases may be disfigured because of prolonged scratching—some suffer an insatiable itching—and the resulting infections. ... Kaposi's sarcoma may intensify the effect. The mouth is often lined by a fur-like layer caused by the Candida fungus, which may also affect the teeth. The layer of subcutaneous fat under the skin has mostly disappeared, so that the contours of the bones are clearly demarcated. On the other hand, the appearance of people with Kaposi's sarcoma may be different. The face and other parts of the body are then puffy and swollen odedematously, a condition which may become so serious that the eyes are no longer visible. This is the result of a congestion in the lymphatic system, with the vessels no longer able to carry away excess fluids.

> *If not confined to bed, the affected person may move about slowly and with difficulty. The finer movements are no longer fully in control. Facial expressions tend to become rigid, speech becomes slurred, and comprehension slows down. Loss of memory and the often occurring personality changes complete the picture of somebody who has the appearance of having aged several decades within a short space of time.* (1989, p. 31)

A central insight to be gained from such tragic conditions is that we each possess a "biological individuality," a bodily integrity which requires that only very individualized substances are carried within us. It is common knowledge that when we eat meat, for example, our digestive system must break it down completely and rebuild (synthesize) it again in the liver and elsewhere. In this way, nothing of the original animal qualities remain in it. The popular saying "you are what you eat" is false. Lovers of fried chicken don't develop feathers and peck at their food aggressively. More apt is the little "poem" by Walter de la Mare that I like to start my 11th grade block with:

> *It's a very odd thing,*
> *As odd can be,*
> *That whatever Miss T eats,*
> *Turns into Miss T.*

In the activity of our immune system we have a highly individualized, but complementary activity to our normal digestion. What we are witnessing in HIV/AIDS is the decline of the capacity to preserve our "biological individuality." The ability to distinguish "self" and "not self" at a bodily level is gradually lost as the T helper cells grow fewer and fewer.

When discussing this with the students, it is possible to point out—once again—interesting parallels with our inner life. Young people at their age are familiar with a certain kind of challenge that life places before us all again and again: situations where a choice must be made as to whether we are going to "remain true to ourselves" (to our values and moral insights), or are we willing to make compromises in order to be more popular, get a better grade, or achieve success in some form or other. In extreme cases, we even speak of someone "selling their soul," which they do at the cost of their inner substance and integrity. In other words, they are not able to say "no" to certain enticements and let these influence what they do and, thus, who they are. Substance abuse can often lead to this, but it can also be a consequence of the emptiness that follows from the compromising of one's core values. There is much food for thought here, but the teacher must tread lightly so the students do not sense a preachy moralistic tone that could hinder them from coming to their own "individualized" conclusions on this essential aspect of human existence (both biologically and psychologically).

As we near the conclusion of this look at HIV/AIDS, students are moved to hear about the tremendous challenges this illness still provides today for millions of individuals and families: in 2018 there were 38 million HIV infected individuals across the globe, with almost 2 million of those under the age of 15.

We can then speak about the various ways in which the virus is transmitted and how the illness progresses after the infection—from the incubation period all the way to full blown AIDS. Lastly, they want to learn something about the kind of HIV treatment that is available (which does not mean it is readily available and affordable to many). They are often surprised to hear that there is no cure for HIV, and not until the middle to late 1990s was a treatment developed called HAART (highly active antiretroviral therapy) that could seriously inhibit the progression of the illness, even without being a cure. They are interested in any stories a teacher might have to tell—if the teacher is old enough to have them—about the early years when HIV/AIDS was first spreading and there was little insight into it all. As far as treatment goes, I usually tell them how, in the 1990s, my wife and I visited a friend of ours who was HIV positive,[69] thinking this would be the last time we would see him alive. His partner had just died of AIDS, and it looked like he would be going down the same road soon. Then, amazingly, at just that time HAART was introduced (a few months too late for his partner), and as a result our friend still continues to live an active life decades later! (Even so, the treatment is not without difficult side effects, is very expensive, and is not a cure.) In the years ahead, coronavirus disease 2019 (COVID-19) will certainly be an important topic in this context.

---

69 This means—if it hasn't been explained already—of course, that HIV antibodies have been found in that individual's blood, which is an indicator they have been infected and their immune system is fighting it, even if unsuccessfully.

The Respiratory System

After spending so much time discussing a microscopic realm that we cannot experience with ordinary consciousness, it is a relief for the students to turn to the lungs and breathing, which is an aspect of our bodily organization that we can directly observe, at least to some degree. In that spirit, one worthwhile way to begin is by taking a few minutes, during which the students breath quietly and try to observe what is actually taking place. What is moving when and how in the chest area, in the stomach area, etc., by in and out breaths? How does this rhythm compare with that of the heart?

From there we can have them all hold their breath for 30+ seconds, followed by descriptions of what they experienced. These observations make clear that we have some conscious control over our breathing—which is an absolute necessity in order to speak and sing. Nonetheless, there are very definite limits to what we can do. They all know that if your breath is cut off for very long you pass out, and if this were to continue, you would die very quickly.

The students can also try to determine the ratio of breaths to heart beat (pulse). They notice right away that this is not so straightforward: when they focus on finding the pulse, they stop breathing for a moment; once they find it they must try to relax and return to their natural breathing rate. Not so easy. Clearly, our breathing is much more directly influenced by our consciousness than our heart rate is.

We have spoken a great deal about how the immune system protects us from foreign elements; we have spoken about the importance of saying "no" to much of what is around us in order to maintain our "biological identity." What's more, all the food we take in must be broken down completely (digested) before it is absorbed into our bloodstream.

But then we look at breathing and find no such barriers. With each breath we take in the air around us. Yes, beginning with the nasal passages we "prep it" a bit (filter dust particles, warm it and moisturize it), but otherwise we take the air directly into our body. In fact, we are sharing this same air with each other in a closed classroom. What I am now taking in, you breathed out a few minutes ago! We all know what it is like to enter a classroom where 20 students have been for an hour with no windows open—that air is not only no longer "fresh," it has also changed greatly. It is now moister and contains more CO_2, which was created through metabolic processes in their bodily tissues, and less O_2, which was consumed through those same processes.

After this warmup, we take a look at the large drawing of the bronchial tree on the board. It certainly reminds us of a tree, but of one that is upside-down.

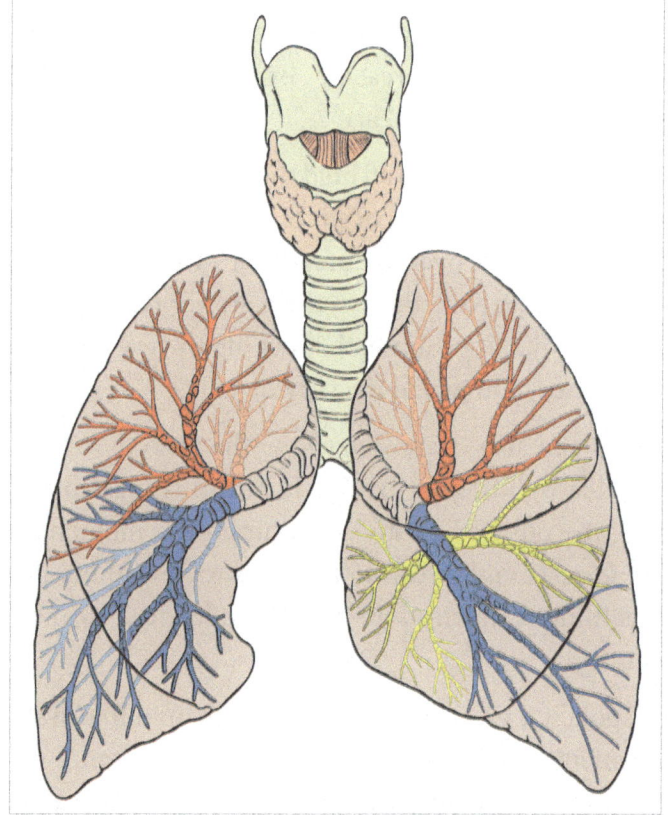

Fig. 2.26 The human lungs.
(Source: Patrick Lynch, medical illustrator, Creative Commons Attribution 2.5 License 2006)

Further discussion determines that this tree is not only downside-up, it is also an outside-in one: It is hollow— filled with air—and has a solid world around it. How about its function relative to a tree? A normal tree, we note— with insights coming from the 9th grade chemistry block—draws CO_2 out of the air, while giving away O_2 to the atmosphere. What does our bronchial tree do? It takes in the O_2 that the trees have given off,[70] and it gives off CO_2 that the trees take in. But with what effect? Our tree uses the O_2 it takes in to "consume" the body around it! That O_2 is used to free up energy by "burning up" (internal respiration) the bodily substances we have taken from the plant world (directly or indirectly) as food through our digestive system.

We are constantly involved in a balancing act between how much we take in through our mouth (eat) and how much we take in through our nose (breathe). Clearly, if we eat less than we burn up, we become thinner and thinner and lose weight, as our upside down, inside-out (hollow) lung tree provides the essential actor (O_2) involved in the consumption (burning up) of our bodily substance. On the other hand, if we eat lots and breathe less (don't get much exercise), our bodily circumference increases and increases with an accompanying weight gain.

It is time to take a closer look at our tree. After gathering observations from the students about what they see in the drawing of the bronchial tree, we can then descend step by step from the top (the base of the tree!) and note key characteristics. At its base (the top) we find the larynx—that we discussed in detail in the 9th grade—sitting atop the trachea. The trachea's mucus- and cilia-lined tubular passageway helps protect against dust particles, and the C-shaped cartilage rings prevent it from collapsing during in-breaths. It divides into the right and left primary bronchi, which also contain cartilage rings and ciliated mucous membranes. The ridge where they separate, the carina, contains a highly sensitive mucous membrane that often triggers a cough reflex. At this branching point, the right branch is steeper and wider than the left. A consequence of this—and a fun fact for the students—is that if they should one day breathe in a button, a bottle cap, or a small diamond, the search for it should be directed down the right branch, not the left!

We have now entered the two asymmetrical lungs—the right has three lobes, the left two. The primary bronchi divide into secondary bronchi, which continue to branch, forming still smaller (tertiary) bronchi, that divide into

---

70 The air we inhale is about 20% oxygen. The air we breathe out contains roughly 15% oxygen. If we do the math, we realize that with each breath we consume about 25 percent of the oxygen we have taken in. That is a lot. Per day it adds up to about 550 liters of oxygen consumed. This is assuming we don't exercise very much. If we do, this number increases greatly.

bronchioles,[71] that continue to branch till they end in the terminal bronchioles, which subdivide into respiratory bronchioles, which divide into several alveolar ducts, around which numerous alveoli are grouped. From the trachea all the way here, our tree has branched about 25 times[72] and ends in about 30,000 terminal branches that "blossom" into some 300 million bubble-like out-pouchings called alveoli!

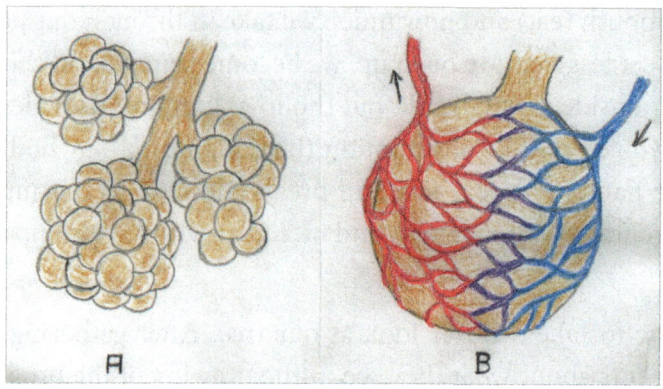

Fig. 2.27 Pulmonary alveoli. **A.** Cluster of alveoli (alveolar sac). **B.** Single alveolus with capillary network.

When grouped together, the aveoli look from the outside like bunches of grapes. And they are truly the fruit of this journey from the base of the tree to the ends of the now microscopic branches. For here, on the outer surface of the alveoli, we find a network of capillaries. The membrane that unites the capillary and alveolar walls (the respiratory membrane) is unbelievably thin: 0.5 micrometers, which is 1/16 of the Ø of an RBC. It allows the O_2 to pass from within the alveoli to the capillaries on their outer surface, and the CO_2 in the capillaries to diffuse inward into the alveoli.

At this point we can ask the students why we have all the tiny grapes instead of a nice big (hollow) peach at the end of the branches? Well, this brings us back to the significance of surface area for the physiological processes of the body (a

---

[71] Along the way the cartilage rings gradually disappear and smooth muscle begins to encircle the lumen in spiral bands. Without the cartilage as support, however, muscle spasms can close off the passageways, which can be life-threatening during an asthma attack. On the other hand, during exercise the relaxation of the smooth muscle dilates the airways and improves lung ventilation.

[72] As we get toward the end these branches, the mucus and cilia fade away and our old friends, the phagocytes, take care of any particles that have gotten that far in.

favorite topic in this block!). If we had peach-like alveoli instead of bunches of grapes, we would not be able to walk at a decent pace without gasping for air due to the lack of sufficient O_2 passing from the alveoli to the capillaries with each breath! As it turns out, however, our 30,000 tiny alveoli create a surface area of about 760 square feet to facilitate gas exchange in the lungs! (This is about the size of a 25 ft. x 30 ft. piece of paper—all of which is folded up into countless tiny alveoli.)

Reflecting on the above, it begins to dawn on the students that when we come closer and closer to the periphery of our tree, the less solid substance there is. That where the respiratory "rubber hits the road" there is nothing there but air-filled pouches that have a boundary that is almost nonexistent at only 0.5 micrometers. This is—in the literal sense of the word—not a very substantial organ!

How does such a frail boundary bring about the movement of air in and out of the lungs? It doesn't! Then how does it happen? That is the next logical question that leads us to a description of the activity of the diaphragm and the intercostal muscles during inhalation and exhalation, as well as the important role that the double-layered pleural membrane plays in this. Total capacity, tidal flow, maximum flow, and residual volume are things that can be discussed in this context.

As we move toward the end of this chapter, we can inquire—as we have regularly done in the block—if there are any noticeable links between our breathing and our inner life. It doesn't take long for several obvious connections

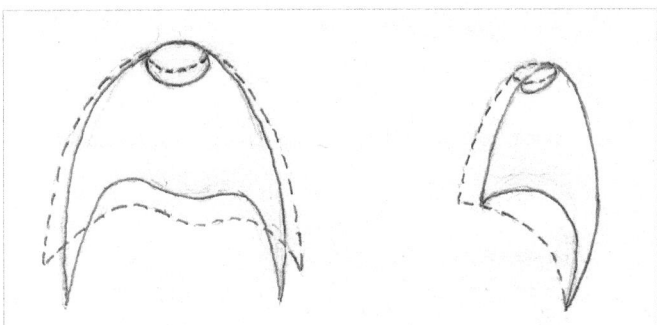

Fig. 2.28 Changes in chest cavity volume during respiration. Dotted line = in breath: intercostal muscles and diaphragm contract. Solid line = out breath: intercostal muscles and diaphragm relax. (After Moerike, Betz & Mergenthaler 2007)

Part II. 10th Grade

to occur to students. Sighing is an easy example, or how, when something scares us, we gasp, and even stop breathing for a few seconds. They remember, too, how it was in the early days of their Waldorf acting careers: how when they appeared on stage they were afraid that—next to their racing hearts—their lungs would freeze up and they would not be able to get any words out of their mouths. The students are delighted when they realize that laughing is actually a sequence of "breath-bursts," while weeping can move between the extremes of strong chest- and diaphragm-moving inward breaths (sobbing) followed by intense outward ones (howling). And then there are those screams that suddenly appear when the grooviest pop star of the decade walks by! In conclusion we often look at the following graphs that show how respiration changes during various emotions.

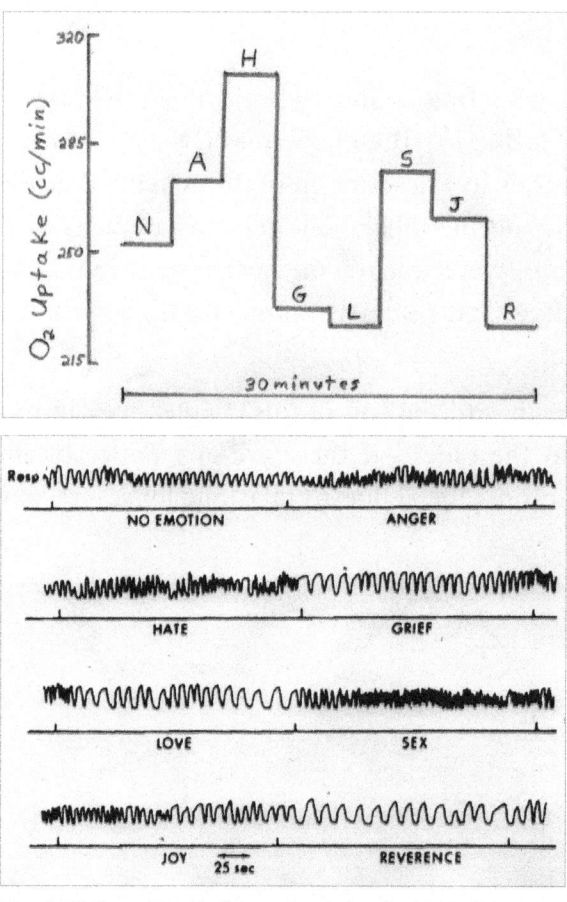

Fig. 2.29 Emotions influence respiration. N = No Emotion. A = Anger. H = Hate. G = Grief. L = Love. S = Sex. J = Joy. R = Reverence. (Modified after Schaefer 1979)

Respiration accelerated during anger and hate. During grief, the respiration has a gasping character, with rest periods at the expiratory end of the cycle. Respiration slows during love and speeds up markedly for sex. It is particularly noteworthy that during reverence there is a marked slowing down of respiration with resting phases at the inspiratory phases of the cycle. (Clynes 1973)

Our study of the respiratory system also includes a brief look at various respiratory disorders, such as asthma, emphysema, pneumonia, tuberculosis, pleurisy (wet and dry) and the common cold.

The Brain

We can approach the next organ in many ways, of course, but as always, we are trying to move the students beyond the deep-seated clichés they bring with them from everyday life. Our goal—beyond giving them an overall grasp of central brain functions—is to widen their horizons regarding the truly dynamic and amazing nature of the brain, while also awakening them to the role they play and responsibility they have in shaping it.

I often start by pointing out a surprising fact: Although it is only a 3½ pound organ, wrinkled-up in appearance and very soft—we can slice it with a butter knife and not feel anything!—that floats within the skull, the human brain is considered to be the most complex "object" in the entire universe!!

That is no small thing! But in what way is it so complex? Just from a numerical standpoint—without even mentioning its amazing way of working—we are dealing with approximately 100 billion nerve cells (neurons),[73] each of which is connected to around 1000 other neurons in an adult brain. Harvard professor John Ratey (2002) did the math for us and says that this allows for approximately 40 quadrillion (40,000,000,000,000,000) possible patterns of connection for an individual brain. Complex and then some!

And then some more, because if we factor in that each connection (synapse) has at least 10 different strengths,[74] then the possible configurations in a single brain rise to an unfathomable number: ten to the trillionth power says Ratey. What's more, the connective patterns in our brains are changing every second in connection with what we do, think and perceive. Taken all together, current understanding, as Ratey points out, views the brain much more as an ecosystem than as a machine.

If there are really so many possible ways these neurons can connect with each other, how do these connections get determined? That, indeed, will be one of our central questions once we lay the ground work with some spatial

---

[73] 30,000 of these neurons would fit on the head of a pin. Placed end-to-end, the neurons in one person's cortex would stretch 100,000 miles, or four times around the earth (Jensen 2015, Korade & Mirniks 2014).

[74] Which is merely a convenient number to use here, since there are over 40 different known neurotransmitters that connect across the synapses.

representations of the brain, including a description of some of its basic functions and key components (synapses, neurotransmitters, glial cells, etc.).

Viewed from above, the brain reminds us of a piece of sea coral that is split down the middle by what is called the longitudinal fissure.

Looking at it from the side, it can be compared (but need not be) with an old, wrinkled boxing glove.

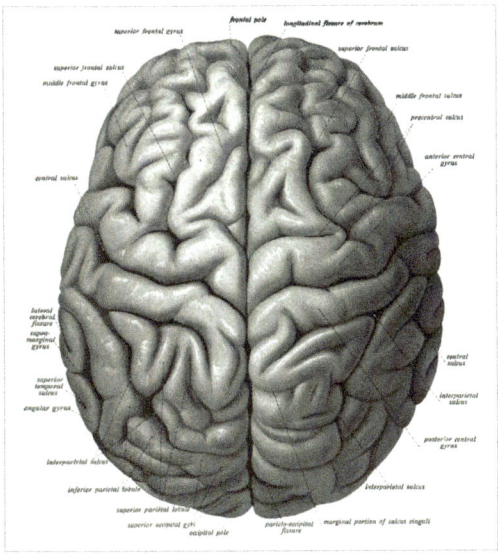

Fig. 2.30 Human brain viewed from above.
(Source: *Sobotta's Textbook and Atlas of Human Anatomy*, 1908)

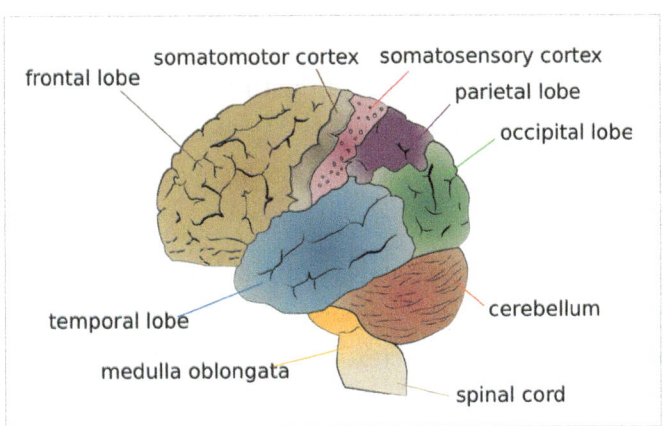

Fig. 2.31 Lateral view of the left cerebral hemisphere and what is located below it.
(Source: http://training.seer.cancer.gov/module_anatomy/unit5_3_nerve_org1_cns.html)

Aided by the larger fissures visible in the drawing, we can distinguish the four lobes of the brain: frontal, parietal, occipital and temporal—the latter of which appears a bit like the thumb of the boxing glove. At the back and below the "boxing glove" we notice the cerebellum, with its very different surface, peeking out.

A frontal section reveals that the longitudinal fissure creates two distinct sides to the brain: the right and left hemispheres.

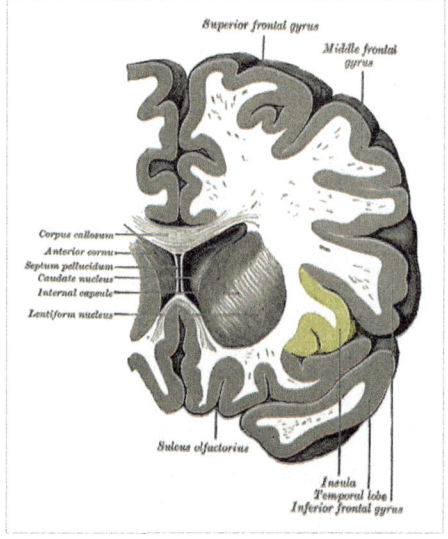

Fig. 2.32 Partial coronal section of the human brain. (Source: File:Gray743.png, 2009)

These are connected by the fibers of the corpus callosum that cross above the paired first and second (lateral) ventricles.

We also notice the many folds[75] of the thin outer layer, the cerebral cortex (*cortex* = bark), that is also referred to as "gray matter." The thickness of this layer is roughly comparable to that of an orange peel relative to the rest of the orange.

Passing upward from the spinal column below and into the "white matter" of each hemisphere are projection fibers, which then fan out (corona radiata) into the projection areas of the cortex.

---

75 At this juncture it is fun to see if any of the students are able to say what the value of so many folds might be for the cortex, in the hope that a recurring theme in this block might jump to mind: an increase in surface area!

The Brain

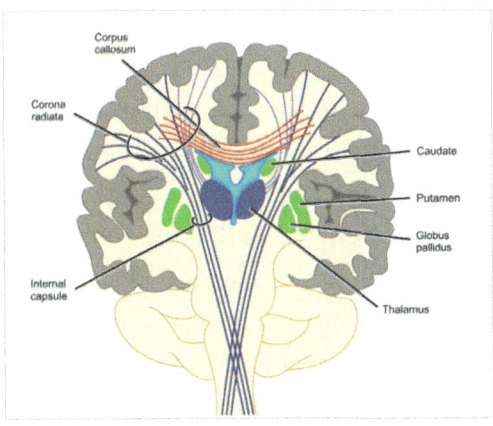

Fig. 2.33 Coronal section with projection fibers.
(Source: Moeller K., Willmes K. and Klein, E. 2015)

With these structural images as a background, we can now move to the amazing fact that all our sensory experiences—by means of which we open up to the world around us (touch, warmth, proprioception, sight, hearing, etc.)—are transmitted to the brain through the projection fibers just mentioned and organized into the various projection areas of the cerebral cortex. In other words, almost everything we experience within our bodies and across its entire outer surface appears to be projected onto the thin surface (the "bark") of one small, wrinkled up organ that is hidden away and floating within the hard bones of the skull.

With the projection fibers connecting directly to their specific areas in the cerebral cortex of the brain, it becomes possible for us to become conscious

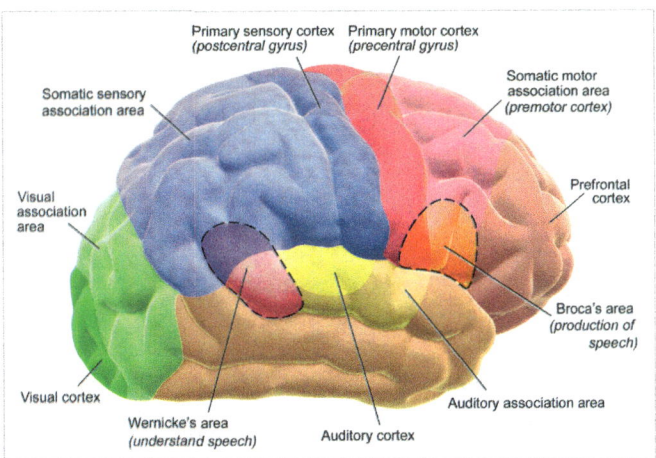

Fig. 2.34 Sensory and motor areas of the cerebral cortex.
(Source: Blausen.com staff [2014]. "Medical gallery of Blausen Medical")

of all the experiences that arise through our various senses. This apparently straightforward organizational principle led to the view—which dominated most of 20th century neuroscience—that the brain is "hardwired." It was formulated succinctly in one sentence by an eminent neuroscientist and Nobel prize recipient Ramon y Cajal: "In the adult (brain) centers the nerve paths are something fixed, ended, and immutable" (cited in Schwarz & Begeley 2002). This view, that brain structures are fixed in form and function, meant that adults are essentially "stuck" with the way they are. Not a happy picture for most people, but one that was widely accepted (and that to teenagers appeared confirmed when they looked at the adults—parents and teachers—around them!). Much more exciting, however, is how radically this view changed as we moved into the 21st century.[76]

One way to introduce the students to today's dynamic picture of the brain is to begin at the beginning with its embryological development. These images

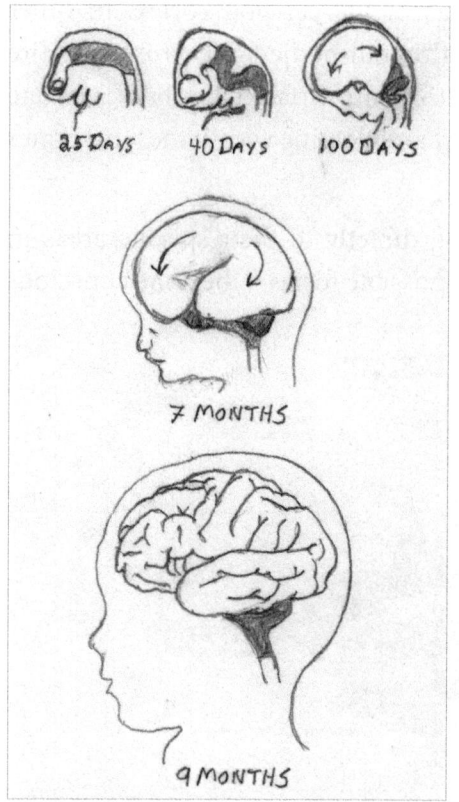

Fig. 2.35 Human embryonic brain development. (Modified after Restak 1995)

---

76   This change has a huge pedagogical significance for students when they learn about their teen years. This should become evident as we move further into the topic.

give a sense for the way in which the brain develops from a tubular form that folds back over itself and gradually develops ridges (gyri) and furrows (sulci). What we cannot see here is how during this process the cellular units of the brain are migrating and developing. By the time of birth, 100 billion or so brain cells (neurons) have been produced. The embryo begins producing neurons at around three weeks, reaches its peak production in the seventh week, and is essentially finished by the eighteenth. If we do the math on this (or let the students do it), we find that the fetal brain is producing around 500,000 neurons (neuroblasts) per minute during its high-production phase—or 250,000 per minute when averaged over all nine months (Ratey 2001, Restak 1995, Schwarz & Begley 2002).

After being produced on the inside of the neural tube, the neurons migrate out to various regions of the brain. Most migrate straight out until they reach the developing cortex. Their further differentiation is determined by where they have migrated to. For example, neurons that migrate into areas where visual input is received differentiate into visual neurons. During their journey, glial cells[77] ("nanny cells") guide and feed them. These two kinds of cells are the primary cell types found in the brain. Once they reach their final location, the neurons grow dendrites and axons, which give them the possibility to communicate with other neurons. Known as "arborization," this process is like a tree growing extra branches and roots. Every sensation, every experience, contributes to the creation of these new neural pathways (Jensen 2015, Ratey 2001).

If things have gone well up to this point, the students should be shouting from their seats: "Teacher, teacher, please tell us more about dendrites and axons, please!" And the teacher will happily meet this request with a simple overview (not included here) of axons, dendrites, neurotransmitters and synapses—while saving myelination for later.

And now comes the exciting part. Each of these neurons, thanks to its many arms and legs (dendrites and terminal branches of the axons) can now "reach out" to other neurons—thereby creating more and more synapses (synaptogenesis)—and through constant communication are able to team up

---

[77] Since the glial cells provide support for and nourish the neurons, they are sometimes referred to as "nanny cells." They do not conduct nerve impulses themselves and are far more numerous than neurons.

Part II. 10th Grade

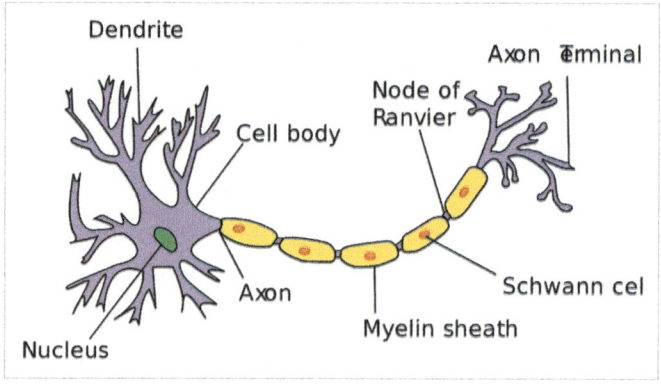

Fig. 2.36 Diagram of a neuron.
(Source "Anatomy and Physiology" by the US National Cancer Institute's Surveillance, Epidemiology and End Results (SEER) Program. GNU Free Documentation License)

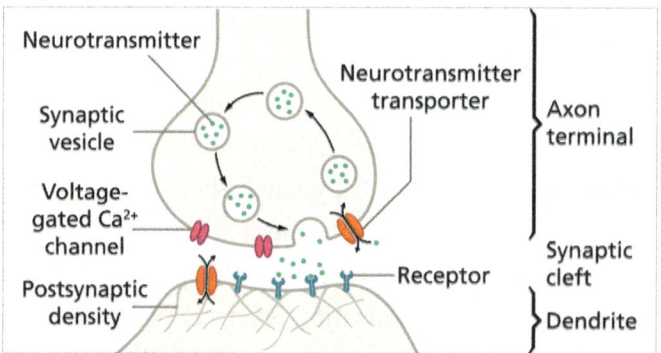

Fig. 2.37 Schematic illustration of a synapse.
(Source: Thomas Splettstoesser (www.scistyle.com). File:SynapseSchematic en.svg. Creative Commons Attribution-Share Alike 4.0 International license)

with them in the most varied ways. We can see how this networking process develops in the first months of childhood.

It is easy to picture how excited young toddlers get: racing around experiencing everything they can get their hands on and their mouths around. And in doing this their neurons are building more and more connections that allow for those discoveries to become learning experiences. Just as we say in everyday life that something makes an impression on us, so at the neurological level an impression is being left behind in the form of new synaptic connections between neurons.

The Brain

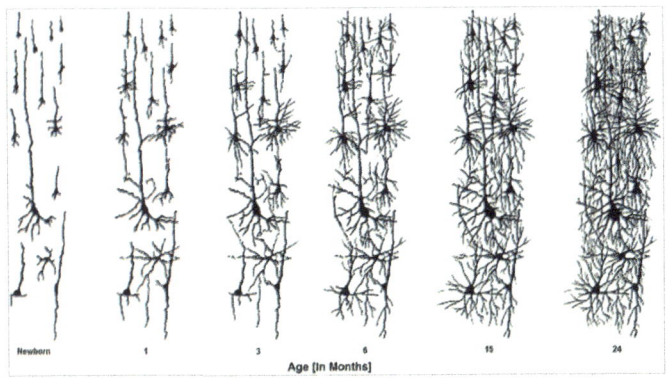

Fig. 2.38 Brain cells developing connections during the first two years of life. (Source: Leisman, G. & Melillo, R. 2015)

My discovery, referred to earlier, that the chocolate-colored soil in my grandmother's garden looked yummy and felt wonderful, but tasted awful, made a big impression on me and on my neural network, as well. Several very different realms of sensory experience came together: Visual (dark brown), tactile (soft, crumbly), and taste (yuk!) impressions were linked together in my inner life in that moment, and for future recognition also for my cerebral cortex through what are known as interneurons (connector neurons, relay neurons, association neurons). These neurons connect various areas of the brain and let us experience as a unity what would otherwise not be linked together. They make up most of what is known as the "white matter" of the brain, which is located below the "gray matter" of the cerebral cortex.

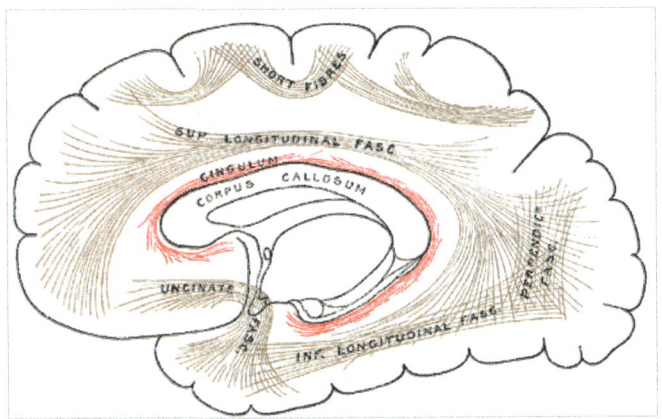

Fig. 2.39 Diagram of interneural (association) fibers connecting different areas of the cortex.
(Source: Gray751_-_Inferior_longitudinal_fasciculus.png)

Part II. 10th Grade

Another attempt to illustrate such interneural networking for students could go something like this: If, prior to the development of the relevant interneurons, I were to see a robin singing on a branch, I would experience the beak opening and closing and would hear singing, but would not be able to relate the one to the other as two sides of the same phenomenon. These two very different sense impressions come to me through very different sense organs and would remain disconnected because the nerve impulses that travel from the two organs travel to different parts of the brain. If however, I attend to what I see and hear with focused attention and think about their relationship, then I grasp what is going on and my brain begins to change. It begins to lay down a connection between these two areas of the brain (the visual and the auditory), forming interneurons (association fibers) that link these two different sensory phenomena together so that soon, with repetition, the connection between them will become strong and therefore obvious to me. My brain will have changed through this learning process and what was unclear to me in the beginning is now a "no brainer" (but should be called a "networked brainer"!).

Cross country skiers are sometimes used to illustrate the process of imprinting the brain through the formation of interneural connections (Bos 2017). Moving across the snow, the skier leaves a track behind. This impression in the snow is made by the movement across it. The next time the skier wants to ski across this section of the mountainside, they will use the track already laid down because it's obvious and it's easier. The more often that person skis along this path, the more consolidated it becomes and the easier and easier it gets—until they can almost "do it in their sleep."[78] (Bos 2017; Kandel, Schwarz & Jessel 2000; Ratey 2002).

On the other hand, if the skier doesn't use that track for a long time, it will snow over and disappear. They should not despair, however, since as humans we can set about making a new track if we so decide and are willing to apply renewed focus and will power to the task. In fact, we can often experience real joy in "breaking new ground" in this way. The phenomenon we are describing here is called "neural plasticity" and has been characterized in different ways through phrases such as "use it or lose it," or with a more neurological flavor:

---

[78] In this context, it is interesting to ask the students if they have ever noticed how little kids love to do the same thing—or hear the same story—over and over without ever tiring of it.

"neurons that fire together wire together." One author writing on this topic offers a short anecdote to make this point: A conductor, who was on the way to his first rehearsal at Carnegie Hall in New York City, was unsure how to get there and asked someone on the street, "How do I get to Carnegie Hall?" The person responded: "Practice, Practice, Practice" (Bos 2017, p. 17).

How the brain develops these deeper "tracks," is our next question. One means that we have already heard about is through the creation of new synaptic connections (synaptogenesis). As we move through childhood, we develop a rich network of interneural fibers between various regions of the brain. When we begin to practice, however, we are focusing on certain connections in particular, which strengthens them, making them firm and fast. At the same time, other connections are being left behind to "snow over." The loss of neural pathways in this way is called "pruning."[79]

To illustrate this we can take our cross country skier and turn them into a mountaineer (soon to be entrepreneur), who is trying to find a way to get between two mountain villages without having to drive all the way around the huge mountain range between them. First they hike various routes in search of the ideal way through. This involves crossing over multiple cliffs and mountain streams. Once the optimal passageway is determined, the other options are left behind and soon fade away (pruning), whereas the chosen trail is widened and evened out, bridges are built over yawning precipices and whitewater rapids so that before long we have a back country road suitable for rugged 4-wheeler adventurists. But hey, why settle for this (thinks the emerging entrepreneur): We could create a big-time connection between the two villages so that they become towns (maybe even cities!) filled with money-laden tourists traveling between them! So they pave the road, turning it into a two-lane highway, then a four-laner—with a European inspired minimum speed limit of 90 mph—and soon what was once a three-hour journey has become just a 20-minute blast over a high mountain pass!

When we practice and develop new capacities, we are building neurological "highways" in our brain. When we try to learn a new language, we find ourselves

---

79 As it turns out, the intense learning that takes place in the first years of life leads to a connective peak of around 15,000 synapses per neuron at age two or three. Thereafter significant pruning begins.

Part II. 10th Grade

in the mountaineer phase of our learning. Every new word and phrase is an obstacle to be overcome, but slowly, slowly—if we stick with it!—we begin to find our way through the confusing landscape, and things begin to make sense. Instead of arduous plodding, one step at a time, we develop a spring in our step and begin to have fun trying to express ourselves and understand others. And, sooner or later, we are racing along reading world literature in a language in which, in the early days, the only word we understood was "Bahnhof."[80]

What happens here at the "brainular" level? How does our mountain story apply to the brain? For one, the precipices and mountain streams that interrupted the trail can be seen as synapses. As it turns out, every new experience causes the synaptic connections between certain neurons to fire and by so doing energizes those synapses. Every time that situation repeats itself a chemical process called "Long-Term Potentiation" (LTP)[81] causes the synapses along that pathway to strengthen, thereby making it easier each time for them to fire in sequence.[82] And, as it turns out, the more motivated the person is, the more their feelings are engaged, the stronger this link becomes through LTP. If this process continues to repeat itself (practice, practice, practice), the neuron chain begins to recruit neighboring neurons to join in. With each repetition the bonds become a bit stronger and more neurons are engaged. Eventually an entire network develops in service of that skill or memory. Learning in this sense exercises the brain and stimulates it, the brain becomes "fit" and develops more capillaries and glial cells (Doidge 2007, Jensen 2015, Le Doux 2002, Ratey 2002).

OK, so we have built very strong bridges (synapses) over creeks, crevices and chasms, but now we need to get the road paved. And here is where a certain type of glial cell—the oligodendrocytes—step up to get the job done. These cells respond to neural activity by wrapping sheaths of whitish, fatty (protein-lipoid)

---

80 If you do not speak German, this is the most important word to know if you ever travel to the German-speaking world: *Bahnhof* = train station.
81 LTP starts with the release of glutamate, an excitatory neurotransmitter that acts as a catalyst and sets off a chain of further reactions that build a larger and stronger synapse with repetition.
82 Neurons subjected to prolonged intense activity show an enlargement of the synaptic end-bulb and a greater number of pre-synaptic terminals. On the other side of the synaptic gap, neurons develop excitatory receptors and show an increase in the number of dendritic branches. The opposite takes place when neurons are inactive. For example, in animals that have lost their eyesight, the visual area of the cerebral cortex becomes thinner (Tortoro & Derrickson 2013).

substance called myelin around the axons of active neurons.[83] The thicker the sheath, the faster the neuron transmits a signal. Mylenated fibers conduct nerve impulses 50–100 times more rapidly than unmylenated ones. The thickness of these myelin sheaths grows in the circuits that are used when we "practice, practice, practice" (Fields 2020, Kranich 2003, Marieb 2012).

All learning and skillful action requires the coordination of many different neurons in diverse regions of the brain. The strength of synaptic connection between the neurons is decisive (LTP), as is the speed of transmission (myelination).

When we are children, this practicing is not a conscious endeavor but comes out of great sympathy for the world around us. A classic example of this is learning to speak our "mother tongue." The language we speak does not come from heredity, but from practicing the language of our surroundings day-in and day-out with immense joy and satisfaction. In the process, the language areas of the brain are shaped and differentiated. As we know, this phenomenal plasticity in regard to language acquisition declines as we grow older, although we still maintain significant potential in this area if we—now consciously—take it up and "practice, practice, practice."

By contrast, it is shocking for the students to hear what happens—or doesn't happen—when children do not find the kind of support they need from a loving environment. This was tragically demonstrated in Romania under the brutal dictatorship of Nicolai Ceaucescu in the 1980s. In an experiment conducted in orphanages, children were left lying most of the day in cribs with high sides. Other than being fed by their "so-called" caretakers, they received essentially no human attention. Without the stimulating interaction and loving contact with other human beings their brain tissues were not stimulated to grow and network with each other, as can be seen from the following brain scans of two three-year-old children. The image on the left is from a healthy child, the image on the right from one of the Rumanian orphans.

The brains of the children who were not able to engage the world actively did not grow and develop as they normally would. Even from such vague images it is evident how the hollow areas of the brain—the ventricles in the center and

---

83 Only the axons become mylenated.

Part II. 10th Grade

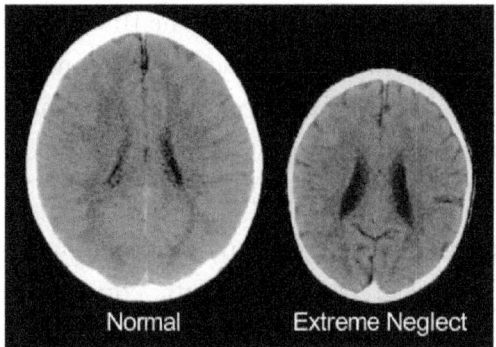

Fig. 2.40 Brain scans of two three-year-old toddlers reveal the effects of neglect. Left image: brain scan of a normal child. Right image: brain scan of a Rumanian orphan.
(Source: Perry, B. (2002) Childhood Experience and the Expression of Genetic Potential: What Childhood Neglect Tells Us about Nature and Nurture. *Brain and Mind*, Vol. 3, 04/01, pp. 79–100)

the sulci in the periphery—are more prominent than in a "well exercised" brain. More detailed examinations showed not only an atrophy of the cerebral cortex (gray matter), but fewer connections linking different parts of the cortex, which is the foundation for rapid and integrated thinking. The saying "use it or lose it" takes on a truly tragic meaning for these children (Begley 2007, Bos 2017, Davis 2020, Perry 2002, Restak 2010).

Building on this, it is important for the 10th graders to learn how significant their current phase of life is in regard to brain development, to learn that they are now in the driver's seat (in more ways than one) and determine to a large extent the direction that that development will take.

Until the late 1990s, it was thought that the great wave of synaptic sprouting and subsequent pruning that characterizes the first years of childhood was over and done with after those early days of dawning capacities. But then, just as we were ready to enter the 21st century, researchers rocked the world of neuroscience with the discovery that just before puberty, a second wave of synaptic sprouting occurs. For example, at UCLA's Lab of Neuroimaging, neuroscientists discovered that the frontal lobes—the area of the brain involved in "executive functioning" (emotional regulation, self control, judgment, organizing and planning)—change noticeably during adolescence. At age ten to twelve (girls usually a bit sooner than boys) the frontal lobes start growing at a pace much like they did during fetal development.

But later, when teens move into their twenties, some areas begin to shrink again, as pruning trims them back into a well-organized, efficient circuitry. So, to the surprise of neuroscientists everywhere, it turned out that rather than being essentially formed and almost finished at age ten or so, the brain of teens and twenty-somethings is an ongoing "construction site"! This second wave of synaptic sprouting and vigorous cortical pruning plays a huge role in the shaping of one's biography and is not confined to the frontal lobes. As Jay Giedd of the National Institute of Mental Health put it:

> *Teens have the power to determine their own brain development, to determine which connections survive and which don't, by whether they do art, or music, or sports, or videogames.* (Schwarz & Begley 2002, p. 129)

Or in the words of another expert, Dr. Frances Jensen, professor and chair of the Department of Neurology at the University of Pennsylvania's Perlman School of Medicine:

> *Everything we do, think, say, and feel influences the development of our most precious organ. ... Our brains, in essence, are self-built... are shaped—landscaped if you will—by the individual's particular experiences. ... Thinking, planning, learning, acting—all influence the brain's physical structure and functional organization.* (2015, p. 58)[84]

Returning to the frontal lobes (the pre-frontal cortex, in particular), their development during adolescence is so central because of the integrative role they play in connecting and coordinating many different inputs. The pruning down of connections and the myelination of those that remain allows for more efficient and specialized coordination of information. The emerging adolescents begin to explore the deeper meaning of life, of social relationships, and also to recognize themselves as the authors of their own ideas. They can think in abstract and conceptual ways that preteens cannot even imagine. They have the possibility to approach problems in creative new ways, to envision a world that is "not yet" but "could be"—and that they feel (think) "should be" (Siegel 2013).

---

84   Since Jensen is referring to humans in general, not just adolescents, she goes on to point out that research has shown that adult brains can be remodeled, too, just not as easily as during childhood and adolescence. One example she brings that the students might want to take home with them—to tease parents and grandparents—is that studies show that learning to tango improves the ability of senior citizens between the ages of 68 and 91 to switch between cognitive tasks (i.e., to become flexible and get "unstuck")!

Part II. 10th Grade

An important takeaway for the 10th graders from all that has just been discussed is well-summarized by Richard Restak, professor of neurology at George Washington University:

If the adolescent is encouraged to concentrate on music, math, or sports, for instance, the brain will incorporate these activities in the form of neuronal circuits. If the teenager, in contrast, spends the day "hanging out" or mindlessly gossiping on a cell phone, the brain will fashion circuits for these activities as well. In essence, adolescents choose the brain cells and circuits that will survive on the basis of the activities they engage in. (2009, p. 17)

Studies have determined that individuals who learn to play a musical instrument, for example, benefit from better communication between the hemispheres. Piano playing exercises the entire brain, which allows other nerve impulses to fly faster and be read more accurately, which has a significant impact on a person's mental acuity. The musician is constantly adjusting decisions on tone, tempo, rhythm, phrasing and style, which trains the brain to become extremely good at organizing and conducting various activities at once. Dedicated practice of such orchestration can have a lifelong payoff in attention and other skills (Schwarz & Begley 2002).

Creative and artistic individuals, in general, have been found to have higher levels of interhemispheric communication. Many studies have shown that training in the arts leads to a higher degree of cortical arousal and extends to good learning in other areas.

Brain changes occur in all areas of study and occupations. This phenomenon was first discovered when it was noticed how cab drivers who had to memorize the complex street patterns of London had a significantly larger hippocampus—that part of the brain involved in spatial visualization and navigation—than normal. Something similar holds true for any specialized occupation. Surgeons show greater activation in the hand area of the cortex than doctors who are not surgeons, for example (Doidge 2007, Ratey 2002, Restak 2009). The same can be said for all of us:

We create new patterns of neuronal organization according to what we see, what we do, what we imagine, and most of all, what we learn.

Learning something new involves establishing a path way within the brain made up of millions of brain cells. As we learn more, these pathways increase in complexity—a process similar to the branching of a tree as it grows. Thanks to its plasticity, the brain can be thought of as a tree of knowledge. When in full bloom, a tree blossoms: Roots give off branches, twigs, and leaves. Similarly, learning increases interaction within the brain with more and more other neurons establishing fuller and richer circuits. But if learning stops, the brain, like a tree losing the luxuriant structure seen at full bloom, reverts to a state corresponding to that of a tree in winter. (Restak 2009, p. 8)

Concluding comment about additional possibilities

In a block such as 10th grade biology, only so much material can be covered. What I have addressed here in regard to the brain is limited in its scope, although it focuses on, in my view, a very essential topic for students in this phase of life. If sufficient time is available, there are many other interesting aspects of the brain and nervous system that can also be explored. For example:

- The fascinating theme of the two cerebral hemispheres, for which Iain McGilchrist's *The Master and His Emissary* provides an outstanding resource.

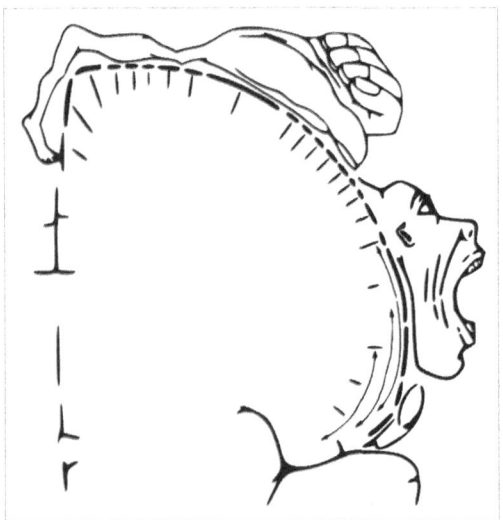

Fig. 2.41 The sensory homunculus.
(Source: File:Homunculus-ja.png After Penfield and Rasmussen (1950), *The Cerebral Cortex of Man*)

Part II. 10th Grade

- The centered versus decentralized nature of the central and peripheral (autonomic) nervous systems—with the spinal cord mediating between them—and their corresponding levels of consciousness.
- The exciting discovery of mirror neurons and their significance; Christian Keyser's *The Emphatic Brain* contains much interesting material on this topic.
- The pivotal role that movement plays in conditioning the brain and optimizing its activity is described in depth by John Ratey in *Spark - the Revolutionary New Science of Exercise and the Brain*.
- The relative proportion of brain tissue allocated to different areas of the body in the "sensory homunculus" is also an interesting topic. (For example, the thumb area grows larger in those individuals who text a lot.)
- And much more.

The Digestive System

A good warmup to this topic is to ask the class to compare hunger and thirst, to describe their similarities and differences: "Where" do we experience them? Does one "feel" more urgent than the other? (The answer to the latter question tells us something about the significance of the "fluid organization" in the life of the organism as a whole.)

The Mouth and Esophagus

Another warm-up question can be: How far does our consciousness extend into the digestive tract? If logistically realistic, it is fun to let the class carefully observe the processes of chewing and swallowing for a few minutes. There are lots of ways to do this; for example, give each student a large scoop of chips and let them experience—with their eyes closed—how chewing and salivating unfold. The role that the tongue plays in all this is also important to note. They should chew slowly and carefully observe what happens. I usually ask them to delay swallowing as long as possible so they notice how some of the liquidy-mass manages to just slip away, but also how at some point the need to swallow becomes urgent. This experiment can be repeated with bread and compared to chip-eating. (Ice cream might even serve as a topper-offer!)

Less appetizing, but also valuable, is to repeat the above process, but to "eject" the bolus out onto a plate about 2/3 of the way through. This allows us to observe more carefully what has already happened to the food at this very early stage of the huge transformation process it will be going through once it disappears on its long journey into a passageway where our consciousness cannot accompany it. We can, however, discuss the little bit we know about the next phase it will be going through based on our experiences of vomit and some of its rather penetrating qualities. They can also try to observe how long it takes after a swallow for the bolus to reach the stomach. (Homework: Pay close attention to how a really full stomach feels and "behaves.")

We have observed (more or less exactly) where wakeful consciousness ends, although there are also subsidiary perceptions like satisfaction, fullness, or even indigestion, that we may experience strongly, albeit less clearly nuanced than in the mouth. At the end of the long path that our food takes, we "wake up"

Part II. 10th Grade

again and realize with varying degrees of urgency the need to expel whatever remains, even though we know nothing about what it went through between tongue and toilet.

From our study of the immune system, it is clear to the class that since foods are foreign substances to our body, we cannot simply take them in without endangering our "biological individuality." If something is foreign then it must be overcome before being "allowed in." What that entails—how extreme that deconstruction process is—is something that the students are about to learn. The need for pre-intake changes before the food enters the bloodstream allows us to point out what would otherwise sound very strange, for our digestive tract—which extends from mouth to anus—is actually a tubular segment of the outer world enclosed within our body! This tube, which runs right through us for 30-some feet, is still part of the outside world. In this organism-external space—which is located internally—the food we imbibe is radically taken apart in preparation for entering the body and before becoming part of our own human substance. I remind the students at this point of the little ditty by Walter de la Mare, referred to when we were discussing the immune system.

> *It's a very odd thing,*
> *As odd can be,*
> *That whatever Miss T eats,*
> *Turns into Miss T.*

Before we take the students on this "hidden journey," the important question for the teacher—as usual—is: What is really essential here? How can we give the students a living picture of human digestion without getting bogged down in too many anatomical and biochemical details, of which there are plenty if we choose to indulge ourselves. There is no "right way" to do this, of course, and it can vary from year to year, particularly if some aspect currently sparks the teacher's interest, so that that enthusiasm can radiate out to the students and subtly awaken their appetites for the topic. I have tended in recent years to move in a more cursory manner through some details of the digestive process in the stomach and small intestine in order to focus more than is often done on recent insights into the huge role that the "microbiome" plays in the health of the human being. I think this topic is particularly important in a time when the fear of bacteria and viruses has become so prominent—even more so since the

COVID-19 crisis—and needs to be balanced out with a wider perspective on the significance of microorganisms.

So what happens in the mouth? Our eating observations showed us that by chewing the food and permeating it with saliva, the substance is so changed that it gradually flows toward the back of the mouth and then disappears from consciousness. Swallowing plays a role in this. We also experience varying degrees of sympathy or antipathy to what we are experiencing.

Not only have we greatly changed the food's consistency—which caused a "yuk" by some when we spit it out to take a closer look—but we are beginning to change it chemically, too. Our saliva is produced by three pairs of salivary glands.[85] They produce watery (serous) fluid that works as a solvent to dissolve foods and which also contains enzymes (salivary amylase) that initiate the breakdown of starch. Another form of saliva is a thick, stringy mucous that helps lubricate the throat (pharynx) to facilitate swallowing. The quality of the saliva changes depending on what we have taken in. If we have something insoluble like quartz pebbles in our mouth, then our saliva will be just water with no enzymes, whereas it becomes enzyme-rich if we put actual food in our mouth. It is also true, however, that just the sight or smell of food—a hot fudge sundae, for example—can get the saliva moving. For some individuals, the mere thought of such a family-favorite triggers major saliva overflow. On a normal day, however, the salivary glands will contribute only about 1–1½ liters to this process. And so begins the first small step on "a long and winding road"[86] (Kranich 2003, Norris 2011, Tortoro & Derrickson 2013).

After a typical "chew time" of 10–20 seconds,[87] the food (bolus) enters the circa 25 cm long esophagus and 4–8 seconds later finds itself in the stomach. (Very soft foods and fluids take only about one second.) This rapid journey is aided by pulsing waves of muscular contraction and relaxation (peristalsis) that pass the food down the esophagus.

---

85 These can be considered in more detail if time is available.
86 Some of the students may call into question the teacher's hypothesis that the Beatles were actually singing about the digestive tract (even if unconsciously) when they produced their famous ballad "The Long and Winding Road." There is, however, no harm in disagreement if one is dealing with art history and not the life sciences.
87 We will have timed this already during our earlier chew exercises.

The esophagus and the entire digestive tract that follows is lined with a mucous membrane that, through its secretions, keeps everything moist, soft, and slippery. In places it also contains other tissues and cells that emit substances into the moving mix. In sum, about 7 fluid liters are secreted into the lumen per day. This process is so dynamic that the stomach lining (the epithelium) must be renewed every two or three days, and the walls of the intestines must be replaced roughly every week!

The rapid tempo that brought food to the stomach ends there, for the rest of the journey (almost 30 feet worth) will last at least another 16 hours. During that time all the food we have taken in at one meal has either been absorbed as nutritional building blocks, or expelled as excrement.

The Stomach

As a warmup to the stomach, I ask the students to point to where it is. Most of them will point in the "tummy/belly-button" area, and are surprised to learn it is higher, roughly at the edge of the rib cage on the left side. From there I like to unveil a drawing on the board that shows the entire digestive tract. This provides them with a spatial framework for our further considerations and will—we hope—whet the student's appetites (get the juices flowing) for what is to come in the days ahead.

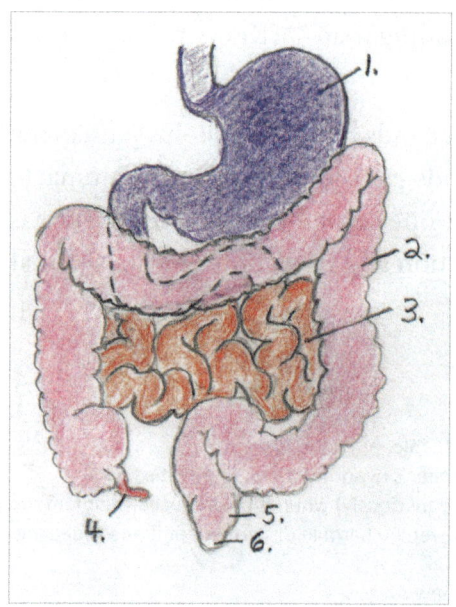

Fig. 2.42 Human digestive tract below the esophagus. 1. Stomach. 2. Large intestine. 3. Small intestine. 4. Appendix. 5. Rectum. 6. Anus.

The Digestive System

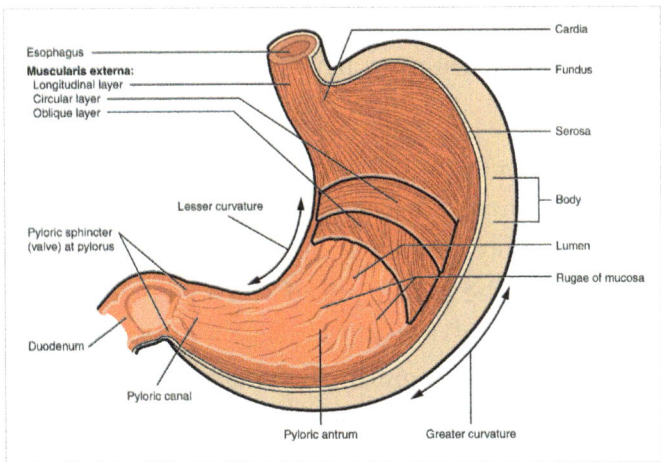

Fig. 2.43 The human stomach.
(Source: OpenStax, College Anatomy & Physiology, Connexions Web site http://cnx.org/content/col11496/1.6/)

Following this preliminary overview of the digestive tract, a more detailed drawing of the stomach emphasizes: a) the three layers of smooth muscle, and b) the folds (rugae) of the stomach's inner lining.

These two aspects of the stomach's anatomy allow us to connect it to experiences—dim as they may be—that most students are familiar with. We discuss with them how the more we eat the more the stomach fills up, although remarkably, there does not appear to be a fixed quantity that equals "full." As it turns out, when the stomach begins to fill, the elastic folds in the lining (the rugae) seen in the drawing begin to stretch out more and more, thereby expanding the volume of the stomach's interior. They keep on doing so until, as we often say, our stomach is "ready to burst." An empty stomach has a capacity of around 2.5 ounces, whereas a full stomach can stretch to hold about a gallon of substance. Such explanations, though not earthshaking in themselves, resonate with the students because they shed light on a realm of their experience that up until now has been very faint, and for which they have had no basis for understanding.

Another dim echo of our stomach's activity that many students have experienced at some time or another is the so-called "churning" of the stomach. We can now relate this to the three layers of smooth muscle contained in our drawing. We explain how these layers of smooth (not voluntary) muscle—the longitudinal, the circular, and the oblique—all contract in different directions.

If we try to imagine these "crisscrossing" contraction patterns and even try to reproduce them crudely with our hands, we can get a sense for how such a churning comes about.

These mixing waves begin to pass over the food soon after it has entered the stomach. They churn and macerate the bolus while mixing it with acidic gastric juice (2_3 liters/day), which eventually turns it into a soupy, oatmeal-like paste called chyme. The gastric juice (which contains mucus, the enzyme pepsin, and hydrochloric acid) not only begins the digestion of proteins, but also activates lipase enzymes that begin the breakdown of fats. Through its strong acidity, it also eliminates pathogens and toxins in food. The mixing waves—with a frequency of every 15–20 seconds—intensify the closer we come to the pyloric antrum of the stomach.

True peristalsis begins only in the stomach's pylorus section, which has a normal two-layered musculature. This means that a true digestive rhythm only begins in the pylorus section, where the food-churning phase ends and the chyme is moved in small portions (of about 3ml) toward the pyloric sphincter, through which it is then transferred into the duodenum. Through this rhythmical transfer from the stomach into the duodenum, the random timing of food intake is balanced out to some degree. Normally, within 2–4 hours after a meal, the stomach has passed its contents on to the duodenum. Carbohydrate-rich foods spend the least time in the stomach, high-protein foods stay somewhat longer, with fat-laden foods being the last to leave (Schmidt 1975, Tortora & Derrickson 2013, Van de Graaff & Fox 1998).

In my experience, a summary of the stomach's activity such as the one above is sufficient if the teacher wants to spend more time on other aspects of the digestive tract, such as the microbiome. But it is certainly possible to investigate additional phenomena relating to the stomach and its functioning. (How does it keep from digesting its own walls, for example?)

The Small Intestine

Reviewing briefly with the students, we remind ourselves that so far the food has been chewed, warmed, salivated, permeated with enzymes and bathed in a highly acidic churning tube. Already in the mouth the first steps to carbohydrate break down began, and in the stomach the first steps in the breakdown of proteins and fats took place. So now, hours later, as we enter the first section of the small intestine (the duodenum),[88] we are still only about 1/5 of the way down the "long and winding road" and much still remains to be done. To the student's surprise, the most intensive digestive process of all is only now beginning, and it is going to take place in the first 10 inches (!) of the small intestine, in the duodenum.[89]

In the spirit of not getting lost in the details, I usually give a short overview of the central processes that take place in the small intestine, in particular, the duodenum, which can be easily expanded upon in this or that direction from year to year, depending on the class and the time available.

Looking again at our drawing of the digestive tract (which I keep on the board until we have completed our study of the digestion), we try to follow its winding path from the stomach until it expands into the large intestine (colon) and ends with the rectum/anus. When this is clear, we shift our attention to several organs that hover around the alimentary canal and which—we will soon learn—make important contributions to its workings.

Building on this spatial clarification, we can describe how all three types of nutrients will now be broken down further—all the way to their basic components—in this section of the digestive tract. This task is accomplished with the help of the pancreas and the liver. The main exit duct of the pancreas empties into the duodenum and provides a wide variety of digestive enzymes. The most central of these are 1) amylase, which digests starch; 2) trypsin, which digests protein; and 3) lipase, which breaks down fats. The so-called "brush border enzymes"[90] come directly from the internal walls of the duodenum and work together with the pancreatic enzymes—in some cases by activating them—in breaking down carbohydrates and proteins. The liver is also an important player here. It takes part in fat digestion by secreting bile, which flows into the small intestine and emulsifies fats, which increases their surface area so they can

---

88   The three main regions of the small intestine are the duodenum, the jejunum, and the ileum.
89   Or, as it is called in German, "the 12 finger intestine," which comes from its being about as long as 12 fingers placed side-by-side.
90   Such as lactase, which breaks down milk sugar into two simple sugars, glucose and galactose. Lactose intolerance occurs when the small intestine doesn't produce enough lactase to digest the milk sugar.

be broken down by enzymes (Rohen 2007, Tortora & Derrickson 2013, Van De Graaff & Fox 1998).

A summary table of all the fluids that have contributed to the digestion process on a typical day[91] (Marieb & Hoehn 2012, Tortoro & Derrickson 2013) can help give the students an overview of what has flowed into that long twisting tube that is enclosed in the body. The 8–12.5 liters prepare the food we ate (hours ago) to be absorbed through the intestinal walls and into the body!

Daily Volume of Fluids

*Swallowed (Ingested)
and Secreted into the Digestive Tract*

Mouth	1–1.5 liters saliva
	2–3 liters liquid swallowed (ingested)
Stomach	2–3 liters gastric juice
Liver	1 liter bile
Pancreas	1–2 liters digestive enzymes
Intestines	1 liter intestinal juice
Total	**8–12.5 liters fluid intake**

It is usually important at this juncture to step back and reflect on what we have been witnessing. In many ways, it has been a process of destruction! What we called food at the outset has been totally taken apart by our digestive system *before* it is allowed to enter the bloodstream. It is important to note, too, that the transformation it undergoes is not one that would take place in the outer world.

Something else happens in the winding tube that prevents the types of catabolism found in outer nature. Instead, proteins are broken down into amino acids, carbohydrates into di- and monosaccharides, and fats into glycerins and fatty acids. These transformations are achieved with the help of the enzymes and other secretions listed above, in collaboration with various forms of intestinal microbiota (that we will soon explore in more detail).

---

91 There is, of course, no typical day, and there can be great variations based on body size, diet, etc.

A few drawings will help the students picture the huge next step in the digestive process—"the breakthrough," if you will—when the broken-down nutrients are finally absorbed into the organism. Several references were made previously to the mucous membrane that lines the digestive tract, but now we go further and take a closer look at the walls of the small intestine in order to determine how their anatomy supports the physiological activities they are involved in.

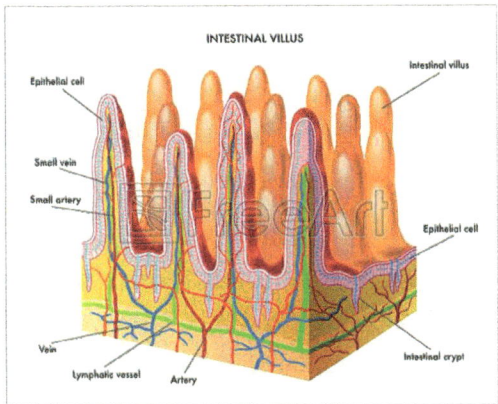

Fig. 2.44 Intestinal villi (singular: villus): the small, finger-like projections that extend into the lumen of the small intestine. Each villus is approximately 0.5–1.6 mm in length and has many microvilli projecting from its epithelium, which collectively form the "brush border." Each of these microvilli is around 1000 times shorter than a single villus. The intestinal villi themselves are much smaller than any of the circular folds in the intestine from which they emerge.
(Source: Free Intestinal Villus Art Prints and Artwork | FreeArt)

Close observation of the drawing above reveals that the walls of the small intestine contain folds (microvilli) upon folds (villi) upon folds (plicae circulars). The students should immediately realize—we hope—that such folding increases enormously the surface area available for absorption of the semi-liquid chyme. These folds upon folds upon folds enlarge the surface area of the small intestine more than 600 times, to about the size of a tennis court (2700 sq. ft.). The villi are finger-like projections that give the surface a velvety texture. The villi are the chief absorptive element in the mucosa and each contains a dense capillary network and a wide lymph capillary, the lacteal. The broken-down proteins and carbohydrates are taken up by the capillaries, the digested fats by the lacteal.

The villi shorten and lengthen alternately like a tongue, tasting and absorbing the chyme that is ready to be taken in. This entire process is supported and enhanced by rhythmical contractions (approx. 12/min.) of the circular muscle fibers around the small intestine, which causes the chyme to slosh back and forth and brings it into more intense contact with the intestinal walls. About 90% of all nutrient absorption takes place in the small intestine, the other 10%

in the stomach and large intestine. Undigested and unabsorbed material then passes on to the large intestine.

The Large Intestine (Colon)

Looking once again at the overview drawing of the digestive system, we see the transition from the ileum into the first portion of the colon, the cecum. It then rises in the ascending colon, moves horizontally in the transverse colon, downward in the descending colon and angles inward in the sigmoid colon, which terminates at the rectum—the last inch of which is referred to as the anal canal. It has a total length of approximately 5 ft., with the name "large intestine" coming—as the students can "see"—not from its length but from its 2.5 inch diameter, which is noticeably larger than that of the "small" intestine.

The chyme that moves into the cecum and ascending colon has a quite thin consistency and is practically free of substances with nutritional value. It cannot be excreted yet—which would generate a condition similar to permanent diarrhea—without creating serious issues for the circulatory system. Instead, it remains in the colon for up to 12 hours, during which 80–90% of the remaining water is absorbed, as well as salts and some vitamins.

The final stage of digestion takes place through the activity of bacteria that live in the colon. They prepare the chyme for elimination by breaking down protein residues and any remaining carbohydrates, which releases hydrogen, carbon dioxide and methane gases. These gases are referred to as "flatus," which, when they become excessive, leads to "flatulence." (The students have a variety of other names for this condition.) The colon bacteria are active in other ways too, ones which have only recently been discovered and which will be focused on in the next section.

By the end of its time in the large intestine, the chyme has become solid or semisolid through water absorption and is now known as "feces." Peristaltic movements push the fecal material to the sigmoid colon and then into the rectum. Distention of the rectal walls stimulates stretch receptors that trigger a defecation reflex, which in infants leads to automatic emptying of the rectum. As they grow older they learn—with help from their parents—to voluntarily control this process (toilet training!).

The Microbiome

When we have reached this point in our description of the colon and what it does, I like to tell the students how "back in the good old days"—in the first decade of the 21st century and before—what we have just covered, from mouth to anus, would have been about it for this block and we would then have moved on to the liver and kidneys. But alas, I explain, science does not stand still. Indeed, around the turn of the century, rumblings of a whole new perspective on human physiology (and beyond) could be heard in biology departments around the world. Symptomatic of the questions that were arising was the launch of the NIH's "Human Microbiome Project" (HMP) in 2005 and the "European Metagenomics of the Human Intestinal Tract" (MetaHIT) project in 2008. More than seven additional programs were also enacted to study the human microbiome at that time.[92]

Trying to understand the human organism in light of the microbiome has awakened a much more ecological approach to human biology than heretofore. Although ecological perspectives for understanding nature in general are a central theme of an 11th grade biology block, I think it worthwhile to touch briefly on this perspective in the 10th grade where it pertains to the human body.

Ecologists look at the relationships between organisms and their environment. Understanding a tropical forest, for example, means seeing how soil, water, sunlight, microorganisms, plants, and animals coexist in the most varied ways. And just as we have a huge variety of ecosystems on planet earth—from rainforests, deserts and alpine tundra to prairies, swamps and savannas—so too does the human body provide a wide range of habitats for the microorganisms that co-exist in and on it. The difference between life in the warm moist enclosure of the nostrils, on the exposed skin of a forearm, or between the sweat-exuding toes, is huge for microbiota and is reflected in the populations that make these environments their homes.

If the students seem to be drifting away after this introduction to the microbiome, one way to wake up their feeling life—although not necessarily the

---

[92] Moreover, numerous initiatives—such as the "Earth Microbiome Project"—have been initiated to study the microbiome in other realms.

sympathy pole—is to make a small excursion into the realm of what is sometimes referred to as the "socktail." The teacher can tell them how skin-munching bacteria—in particular previbacteria—thrive in the warm, damp, salty milieu between the toes, where they find an unending supply of nutrients in the form of dead skin cells. We have hundreds of millions of these bacteria living happily on our feet as they produce the cheese-like smells—"toe-jam"—that we all love so much. These same bacteria are also used in creating cheeses like Limberger, Muenster, Port-du-Salut and Entrammes. One odiferous byproduct of this activity is butyric acid with its lingering Parmesan scent. I sometimes bring a beaker of it to class to provide the students with an enlivening sniff as we discuss this topic.

In its pioneering study of various human "ecosystems," the HMP identified 15 such habitats in the human body, from which samples were taken.[93] It cannot be our task in this block to examine the amazing diversity of the human microbiome, but a few general findings can be noted before focusing on the gut microbiome. Prior to the HMP, researchers had identified a few hundred species of bacteria connected to the human body. Using new methods recently made available to them, the project was able to identify around 10,000 species of microbes connected to our bodily organization. The study found that each person carries a unique microbiome, but even with individual differences, similar body parts have similar microbes. Armpit microbes in one person are similar to armpit microbes in another. Even when the species vary a bit from person to person, they still perform the same functions.

Turning to the human intestinal tract, we find a diverse and complex community of microbes that play a central role in intestinal functioning and in our health. It is estimated that the human gut contains around 1000 different bacterial species, in addition to the fungi, protozoa and viruses that make it their home. The vast majority of these are found in the colon. In fact, the colon contains the highest microbial density recorded in any habitat on earth, with up to 1012 cells per gram of intestinal content (Ferranti et al. 2014, Shapira 2016)! Their total weight is roughly equivalent to that of the human brain. There is now abundant evidence that the effects of the gut microbiota extend beyond digestive

---

[93] Three additional habitats were studied in women to help determine the role played by vaginal microbes during pregnancy and childbirth.

processes to numerous physiological systems and overall health. It would go too far in this block to expand into such wider connections, but it is important to give the students a quick picture of a few of these confluences.

In a healthy body, the trillions of microbiota form a unique network of supporting processes that play pivitol roles in promoting the smooth daily operations of the organism. When we are healthy, pathogenic and symbiotic microbiota coexist without difficulties. However, if the balance is disturbed—brought about by certain diets, antibiotics, etc.—dysbiosis (microbial imbalance) occurs and these normal interactions are disrupted. As a consequence, the body may become more susceptible to disease. When in balance, the microbiota breakdown potentially toxic foods, stimulate the immune system and synthesize certain vitamins and amino acids. They also break down complex carbohydrates like starches and fibers that have traveled to the large intestine. By fermenting indigestible fibers they create certain fatty acids[94] that play an important role in various bodily functions. In a healthy individual, the microbiota also provide protection from pathogenic organisms that are taken in with food or drink.

The microbiome is a very dynamic network—so dynamic that some researchers say we should think of it as a verb (not a noun) that can fluctuate daily, weekly, and monthly, depending on diet, exercise, medications, and a multitude of other influences from the environment. Of particular significance is its ability to respond to changes from without. Microbiota are amazingly reactive and adaptive, which is central when a new food source or toxin appears in the environment. They help our entire organism to meet a changing world. One recent study found, for example, that a common gut microbe in Japanese people has taken on certain characteristics found in marine bacteria. As a result, the Japanese are able to do something the rest of us cannot do very well: digest seaweed! As one researcher put it: The plasticity of our microbiota gives us a bag of biochemical tricks that enable us to respond to changes in our environment much more quickly and effectively than would otherwise be possible (Chopra & Tanzi 2015, Douglas 2018, Harvard School of Public Health 2020, Pollan 2013).

Our next question is: How do we develop a strong microbiome? One narrative that helps lay the groundwork for this—and that usually interests

---

94 Short-chain fatty acids (SCFA).

the students—describes the work of a young German physician and allergy specialist, Erika von Mutius, who saw an exciting opportunity. She realized that when "the Wall" that separated East and West Germany went down in 1989–1990, an amazing and extremely rare research possibility was presenting itself to someone who was "quick on their feet." Von Mutius saw a research "window" that was opening at that moment and would soon close again: the chance to compare children who were ethnically similar but who, having grown up on different sides of the Wall, had experienced very different ways of life and very different environmental conditions. Whereas the West German cities tended to be wealthier, cleaner and more modern, the Eastern cities were poor and polluted. As an allergy specialist, von Mutius hoped to determine how the differences found on the two sides of the Wall influenced the prevalence of allergies and asthma in the children who had grown up in these contrasting environments.[95]

Before moving on to the findings of von Mutius' research, it is interesting to ask the students what they think the outcome of these investigations will be, and why. Well, to the surprise of von Mutius and many others, it turned out that the children living in the poorer environmental conditions of the East had (proportionately) only half as many allergies, and lower rates of asthma than their peers in tidy, clean and much more affluent West Germany.

How can we make sense of this? The class discusses this, looking for ways of viewing the issue. We return to von Mutius and learn that her research led her to the following conclusion: It is actually good to be exposed to microbes, especially early in life. Put differently, if you limit children's exposure to microbes, they develop over-reactive immune systems and are inclined to develop allergies or asthma. So, if this is the case, we ask the students, what would the ideal conditions for a healthy microbiome be?

After some discussion we learn that von Mutius—now a professor of Pediatrics at the University of Munich—and numerous others explored these questions further. They found that children who grow up on traditional farms—who have early life contact with farm animals and their fodder, and who consume unprocessed cow's milk—have significantly lower risks of asthma and allergies.

---

[95] Allergies and asthma were once rare, but since the 1980s rates of asthma and allergies have tripled, making them the most common chronic (ongoing) disease in American children. Both are indicators of an overreactive immune system.

The more microbes the better! Researchers tallied the number of microbes in the dust from animal stables from different farms and found that farms with the most microbes in their dust showed the lowest allergy and asthma rates (Hirsch 2017, Meyer 2015, von Mutius 2007, von Mutius & Vercelli 2019).

These findings can be related to what we said earlier about how the immune system "goes to school" for many years after birth. This time we ask: Where does the newborn's microbiome come from? In preparation for childbirth, a pregnant mother's body changes in many ways, among these a shift in vaginal bacteria. While many abundant species decrease in number, at the same time certain rare species begin to flourish. One of these rare groups is the genus *Lactobacillus*, whose members can digest lactose, the main sugar in breast milk.

This can lead to a brief mention of childbirth—which may make some students a bit squirmy, but which they enjoy, nonetheless, and usually remember for a long time. During childbirth, babies leave the womb and squeeze through the birth canal. As they do so, each part of their bodies comes into contact with the flexible canal, which coats skin, mouth, nose, etc., with vaginal microbes. In this way, the baby is exposed to a rich mix of motherly microorganisms, including milk-digesting *Lactobacillus*. As one author put it: "Being slathered in vaginal microbes might not seem like much of a treat, but to the newborn, it's a key event" (Yong 2010).

Right after birth, the baby instinctively moves its mouth—which is now full of *Lactobacillus*—toward the mother's breast and begins sucking. In addition to the nutrients in the milk that are for the baby, breast milk is full of additional sugars (oligosaccharides), as well as specific microbes (such as *Bifidobacteria*) that consume those sugars. The oligosaccharides feed these and other important microbes so that they can move into and colonize the baby's gut. Among other things, the microbes play an important role in the development of the baby's immune system. In short: Breastfeeding babies not only obtain nutrition from the mother's milk, they also get a "starter-set" of bacteria and the food that those bacteria need to thrive, multiply, and support the baby's development.

But this is only the beginning. When babies are cuddled, kissed and played with, they are receiving through such contact new microbes. Mom, dad, sibs and neighbor kids all contribute. As time passes, they gather more and more from their surroundings: They put toys—and anything else that will fit—into their mouths, they crawl on the floor, their dog gives them a dripping tongue-full right on the kisser. They also get microbes from what they eat. Breast-fed babies receive many more of the sugar-digesting *Bifidobacteria* than those who are given formula. When the baby starts eating bananas and mashed peas, and drinking carrot juice, other kinds of bacteria flourish whose specialty is breaking down complex carbohydrates from plant foods. By the age of three, a child's microbiome is largely stabilized and similar to an adult's (Dietert 2016, Hirsch 2017, Moise 2017, Yong 2010).

As we near the end of this introduction to the microbiome, I think it is important that the students hear at least briefly about the serious concerns that have arisen in recent years concerning the effects of excessive antibiotic use on the microbiome. From his extensive research as the director of the "Human Microbiome Program" and as Chair of the Department of Medicine at NYU, Dr. Martin Blaser has been instrumental in awakening awareness for the importance of symbiotic relationships residing in the human microbiome and how the overuse of antibiotics weakens that internal ecosystem.[96] Children in the Western world, he points out, receive between 10 and 20 courses of antibiotics, on average, before they turn 18.[97] And those prescribed drugs are not the only antimicrobials that endanger the microbiota. Antibiotic residues are found in milk, meat,[98] and surface water, as well. From "antibacterial" hand sanitizers, to chlorine washes for lettuce, in our daily lives we are influencing our microbiome significantly (Blaser 2014).

Blaser and other researchers are not questioning the value that antibiotics have had historically in helping to deal with many infective diseases and in

---

[96] Although it is important not to paint a too pessimistic picture for the students, the teacher, at least, should know what Blaser sees as symptoms of this weakening of the immune system. In Blaser's (2014, p. 2) own words: "Just within the past few decades, amid all of these medical advances something has gone terribly wrong. In many different ways we appear to be getting sicker. ... We are suffering from a mysterious array of what I call 'modern plagues': obesity, childhood diabetes, asthma, hay fever, food allergies, esophageal reflux and cancer, celiac disease, Crohn's disease, ulcerative colitis, autism, eczema. ... Food allergies are everywhere. A generation ago, peanut allergies were extremely rare. More and more children suffer from immune responses to proteins in foods, not just in nuts, but in milk, eggs, soy, fish, fruits—you name it, someone is allergic to it."

[97] In a recent study, the antibiotic amoxicillin was found to be the most frequently prescribed drug to infants and children in the United States (Dietert 2016, p. 143).

[98] Blaser has also shed light on the problematic use of antibiotics to increase the weight of farm animals.

saving many lives. What they question is the "easy-going" frame of mind around antibiotics that has prevailed among most patients and doctors for the past century. In what is known as "antibiotic overreach," antibiotics are frequently prescribed with the mindset: "Why not just give some antibiotics, it certainly can't hurt." Rodney Dietert, professor of immunotoxicology at Cornell University, describes the situation as follows:

> *The thinking was that inappropriately applied antibiotics had no real downside. No harm, no foul, was the idea. But we now know there is harm and potentially a significant amount of it. Under the old biology, bacteria were generally seen as evil, and widespread killing of our bacteria by antibiotics was no problem. One round of antibiotics can damage your microbiome and cause your entire metabolism to change, along with the interconnected functions of your tissues and organs. ... Non-targeted, beneficial bacteria are also destroyed during antibiotic treatment. ... In the past we didn't know the health ramifications of damaging our microbiome. But we do now.* (2016, pp. 143–144)

An anecdotal story relating to this issue is told by the well-known author Michael Pollan.[99] Pollan explains how in the context of "the American gut project," it was found that his microbiome showed a significantly higher diversity of microorganisms than the typical westerner's. His skin also harbored many bacteria associated with plants, soil and animals. (He is an enthusiastic gardener, composter—worms too!—and he ferments lots of "live-culture" foods teeming with microbes.) Well, as it turns out, soon after this assessment Pollan had to undergo oral surgery and his dentist, as a precaution, put him on a course of the antibiotic Amoxicillin. Within a week the impressive non-Western microbial diversity in his gut had plummeted and now looked much like that of the average American. It showed a spike in bacteroids (much more common in the West) and a disturbing bloom of proteobacteria—a phylum that includes many pathogenetic characters, including *Salmonella* and *E. coli*. What had been a healthy-appearing, diversified gut, now raised concerns among the microbiologists who studied his data.

---

99 Many of the students will be familiar with his book, *The Omnivore's Dilemma*, which is often used in their 9th grade agriculture block.

Summing up: As our understanding of the microbiome increases, it is becoming more and more obvious that the kind of thinking that seeks to pinpoint a culprit when illnesses arise, and then tries to find a straightforward way of getting rid of "the problem," does not do justice to the complex reality of a living organism interwoven with its environment. To understand the dynamic relationships between the body and its microbes requires an ecological way of thinking. The clear distinction that is usually made between body and microbes is less real than we normally imagine it to be. The microbiota play vital roles in the life processes of the organism of which they are a part. They are constantly recreating themselves (think of them as verbs!) within the multi-level and interwoven activities of the whole organism, and in turn, the organism is realizing itself with the help of those active microorganisms.

In conclusion, the words of Michael Pollen can give us an unpretentious and quite reasonable picture of where we stand in relationship to the microbiome at this point in history.

> *It is a striking idea that one of the keys to good health may turn out to involve managing our internal fermentation. But absolute control of the process is too much to hope for. It's a lot more like gardening than governing. The successful gardener has always known you don't need to master the science of the soil, which is yet another hotbed of microbial fermentation, in order to nourish and nurture it. You just need to know what it likes to eat—basically, organic matter—and how, in a general way, to align your interests with the interests of the microbes and the plants. The gardener also discovers that, when pathogens or pests appear, chemical interventions "work," that is, solve the immediate problem, but at a cost to the long-term health of the soil and the whole garden. The drive for absolute control leads to unanticipated forms of disorder.*
>
> *This, it seems to me, is pretty much where we stand today with respect to our microbiomes—our teeming, quasi-wilderness. We don't know a lot, but we probably know enough to begin taking better care of it. We have a pretty good idea of what it likes to eat, and what strong chemicals do to it. We know all we need to know, in other words, to begin, with modesty, to tend the unruly garden within. (2013)*

The word modesty is important here. Many microbiome experts are concerned that profit-seeking individuals are overselling the microbiome in a way that goes far beyond our current insights into its workings. It is being promoted online and in various media as a cure-all for the most varied ailments. In light of this, Jonathan Eisen, a microbiologist at the University of California, Davis, has created the "Overselling the Microbiome Award" that he regularly gives to organizations and individuals he thinks are exaggerating the results of research studies, or making other kinds of misleading claims about the microbiome.

Part II. 10th Grade

The Liver

The venous blood containing products of the digestive process does not enter the general circulation yet, but flows instead into the portal vein (along with blood from the pancreas, spleen and gall bladder), which takes it to the liver.

The pinkish-brown liver is, at around 3½ pounds, the largest gland in the body. It is so soft that when removed it looks like a formless mass. Every minute approximately 1½ liters of blood flows through its sponge-like structure. In fact, the liver contains scarcely more solid substance than the blood itself. (The blood is 78% water, the liver 71%.) In contrast to other tissues, where the water content gradually decreases throughout life, the adult liver maintains lifelong the same high water content it had as a newborn. It also maintains an amazing capacity to regenerate its own tissue. Up to 80% of it can be removed and it will still regenerate—something no other organ can do!

So why does all the blood from the digestive tract flow to the liver before it is "allowed" to enter the general circulation? To give the students an anchor for picturing the activity of the liver, we start with a few drawings showing its location in the body and its inner structure, followed by a description of the blood flow through it.

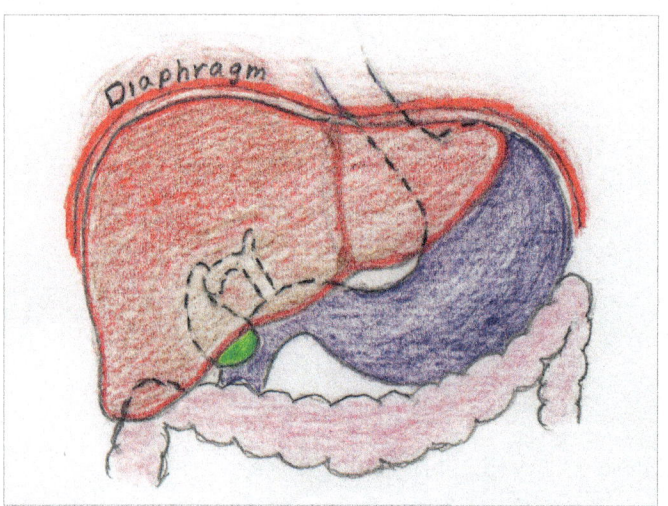

Fig. 2.45 The liver nestled into the dome of the diaphragm above, with the gall bladder (green) peeking out below.

The Liver

The upper surface of the liver is nestled into the dome of the diaphragm, to which it is partially connected. In this way it partakes in the breathing rhythm by rising and sinking with each breath. Its saucer-shaped underside opens itself to the digestive organs below.

We describe for the students how slowly the blood flows—approximately seven-times more slowly than in normal capillaries—from the periphery to the center through the thousands of miniscule hexagonal-shaped lobules that make up the liver.[100]

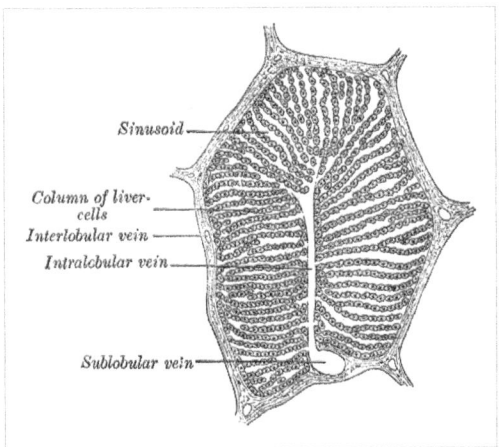

Fig. 2.46 Image of one of the thousands of liver lobules that make up the primary substance of the liver. Blood coming from the portal vein slowly flows from the periphery of the lobule into the central vein, from which it enters the hepatic vein that carries the blood from the liver to the inferior vena cava and on to the heart.
(Source: Henry Gray [1918] *Anatomy of the Human Body*)

As the blood moves though the lobules it interacts intensely with the liver cells themselves. The unusual porosity of the capillaries, which are riddled with openings (fenestrated sinusoids), allows the blood plasma to wash freely around the liver cells. Through this immersion in the blood that is passing through it, the liver is highly sensitive to the blood's make-up as it comes from the portal vein. All of the nutritional substances that were broken down in the gastro-intestinal tract are now re-synthesized, re-enlivened and made into endogenous "human" substances. The liver is also a gatekeeper for foreign matter and toxins. Through its highly-developed capacity for detoxification, contaminants are not allowed to infiltrate the organism (Kranich 2003, Vogel 1979).

In order to bring a "digestible" picture of the liver to the students, it is not possible to describe in detail the many ways that the liver interacts with the

---

100 Their tightly-clustered hexagonal shapes are reminiscent of a beehive.

substance passing through it.[101] Before focusing in on two key aspects, I usually give a brief description showing the immense scope of the liver's activity, such as can be found in the following words of Benninghoff & Goertler (1980, Vol 2, p. 137):

> *All of the substances circulating in the blood…, for example carbohydrates, proteins, and fats, but also hormones, ferments, medications, toxins, and tissue fragments can be changed through intervention of the liver. Which means, they can be broken down or synthesized and in that way made valuable for the organism in a new form—or they can be excreted.*

If we add to this the role that the liver plays in the regulation of water and minerals in the body, and how it regulates the composition and quantity of the circulating blood, we can concur with Otto Wolff (Husemann & Wolff 1987, Vol. 2, p. 205):

> *There is hardly a metabolic activity in which the liver does not participate. It is the place of general substance formation and the origin of human substance. In terms of substance, the process of becoming a human being occurs in the liver.*

The role of the liver is so profound and multifaceted that it is hard for 10th grade students to absorb if we go into too much detail. For that reason, I have found it effective to highlight in particular an aspect of the liver that plays a central role in our everyday life, although we are usually unaware of it. And that is the liver's conversion of blood sugar (glucose) into liver starch (glycogen), and back again.

Blood sugar (glucose) is the main source of energy for all metabolic processes in the body, as well as for muscle contraction and brain activity. It must, however, be regulated so that the glucose concentrations correspond to bodily needs under varying conditions. After a carbohydrate-heavy meal, for example, the liver is able to remove quantities of glucose from the blood flowing

---

101 The liver has a wider variety of functions than any other organ in the body; more than 500 liver functions have been identified (Mosby 2006).

through it. The glucose is then converted into a form of starch (glycogen)[102] through a process known as glycogenesis. If the sugar were allowed to enter the general circulation immediately, then the blood's glucose levels would spike after meals and crash midway between them. We would then live under the direct influence of our digestive system and would swing back and forth (in misery) between states of too much and too little glucose to meet our bodily needs.

Thanks to the activity of the liver, however, it is not only possible to remove glucose from the bloodstream when its levels are too high, but also to do the opposite. When blood sugar levels drop and more is needed somewhere in the organism, the liver begins to transform some of its stored glycogen back into glucose (glycogenolysis) and sends it into the bloodstream.[103]

The students are interested to hear that insulin—a hormone produced in the pancreas that they have heard of in the context of diabetes mellitius— is involved in glycogenesis and when deficient leads to an overabundance of glucose in the blood and corresponding disfunctionalities. Excessive insulin secretion, by contrast, results in abnormally low blood glucose (hypoglycemia), which causes lack of muscular coordination, sweating, weakness, and mental confusion that can lead to the loss of consciousness (diabetic coma)(Bott 2004, Rohen 2007, Rosslenbroich 1994).

What we have been describing is the ongoing, flexible, moment-to-moment regulation of blood sugar by the liver. Behind this stands, however, a larger, more fundamental liver rhythm that integrates us into the 24-hour circadian rhythm of the sun. This liver rhythm is essentially independent of food intake.

We can illustrate this temporal pattern for the students with the help of a simple diagram such as the one below. The graph with the wavy blue and yellow lines shows glucose and glycogen levels in the liver in the course of a 24-hour circadian rhythm.

---

[102] Through their organic chemistry block in 9th grade, the students should have a clear grasp of the significance of sugars, starches, etc., and how they relate to each other.
[103] In classes where interest in such processes is strong, it makes sense to tell them about "gluconeogenesis," as well, which is a process where noncarbohydrate-resources are used to form new glucose when reserves have been depleted.

Part II. 10th Grade

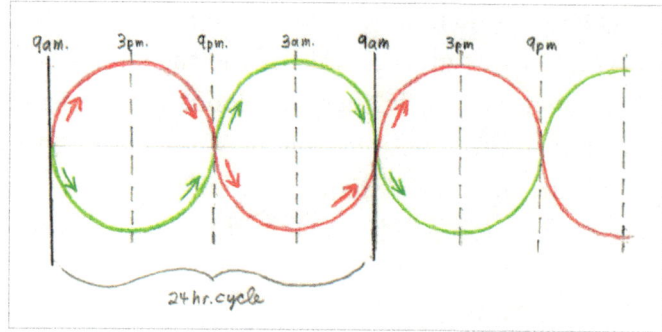

Fig. 2.47 Graph of the 24-hour liver rhythm. The red line indicates how blood sugar (glucose) levels increase and decrease over a 24 hour period, with a maximum around 3pm and a minimum around 3am. When not in the blood, glucose is stored as glycogen in the liver, which is represented by the green line in this graph. When glucose levels are increasing, glycogen levels go down. When glycogen levels rise, glucose is being withdrawn from the blood stream and stored in the liver.

Focusing only on the green section of the diagram above, we relate the time-indicators (3pm, 9pm, 3am, etc.) below the graph with the curving blue line. They show the level of glucose in the blood at different times of day.

- Starting at the leftmost highpoint of the blue curve at 3pm, we see that the sugar content has reached its maximum and decreases downward to the right until 3am, when the glucose levels in the blood are at their lowest (and glycogen levels are at their maximum).
- At 3am glucose levels begin to increase again (since the liver begins once more to transform glycogen into glucose) until blood sugar levels peak at 3pm, after which the whole process reverses and sugar starts to be withdrawn again.
- The opposite pattern is depicted by the yellow line, which represents the glycogen levels stored in the liver. They start to increase at 3pm and max out at 3am. (When one goes up, the other goes down…)

In short: Between 9am and 9pm (the upper half of the blue curve) the presence of sugar prevails over the storage function, which supports our active daytime-activities. At night—from 9pm to 9am—the balance swings toward the energy storage pole (the upper half of the yellow curve). Interestingly, it is not uncommon for people to notice the two transition points in liver activity.

One is when they "run out of juice" in the afternoon around 3pm (when the liver begins to withdraw sugar); the other, when they experience the irritating tendency to wake up at around 3am in the morning (when the liver begins again to send more sugar into the bloodstream).

This 24-hour liver cycle is unknown to many people, even though much research in recent years has shown that the disruption of this circadian rhythm can influence health. How it might be disrupted is a question we can put before the students. Although it may take some time, the discussion eventually leads to people who work night shifts, people who have lifestyles that ignore the normal day-night rhythms, or individuals who frequently travel through different time zones. Many students are familiar with what is known as jet lag, and are interested to learn that, as a general rule, it usually takes about one day per time zone crossed to recover from jet lag (Bott 2004, Ferrell & Chiang 2015, Fonken & Nelson 2014, Kellner et al. 2016, Rosslenbroich 1994).

As mentioned earlier, there is so much more that could be considered in regard to liver activity, but too much is, in my experience, counterproductive. One final aspect that I find it important to cover is the production of bile by the liver. Looking back at our liver lobule drawing, we recall the slow flow of the blood from the periphery to the center of the liver lobules as it washes over the liver cells. Well, as it turns out, there is another flow going on in the liver, this time from the center to the periphery. The liver cells (hepatocytes) perform many functions, one of which is the production of bile. The bile is secreted into small channels (canaliculi) at the boundaries of the liver cells, which combine to form ducts, which carry the bile to the outward to the borders of the lobules and on into larger ducts which flow into the duodenum directly (to emulsify fats as mentioned above) or into the gall bladder for storage between meals. As we learned erstwhile, about 1 liter of bile flows from the liver to the duodenum each day (Marieb & Hoehn 2012).

A good way to round off the study of the liver is with a short depiction of one or two liver-related illnesses that most students have heard of, such as hepatitis and cirrhosis of the liver.

Part II. 10th Grade

The Kidneys

One interesting way to begin the study of the kidneys is to present the students with a medical condition—known worldwide since the earliest days of recorded medicine—that was both a great riddle and a woeful malady. John Wesley (1703–1791), medical author (and founder of Methodism) describes this illness—known as dropsy—in the following way: "a preternatural collection of water in the head, breast, belly, or over all the body. It is attended with a continual thirst. The swelled part pits if you press it with your fingers. The urine is pale and little." A 14th century author simply described it as "a watery disease inflating the body." Well known individuals such as Samuel Johnson and Andrew Jackson suffered from this illness. Jackson, seventh president of the United States and known as a rugged battlefield warrior, died of dropsy. He said of himself in his final days: "I am a blubber of water" (Peitzman 2007).

After introducing the topic with a few anecdotes such as the above, I give the students a handout with the image seen below—but only after preparing them by saying that that which they are about to see is not funny, but represents true human suffering and should be respected in that frame of mind. Nonetheless, I tell them we will attempt to inspect the image carefully to see what we can observe and what conclusions we might draw from what we see.

The illustration of a woman with dropsy comes from a medical text anno 1695. She was an actual patient: 48 years old and a mother of five. The students are asked to make observations. I break them up into small groups of three or

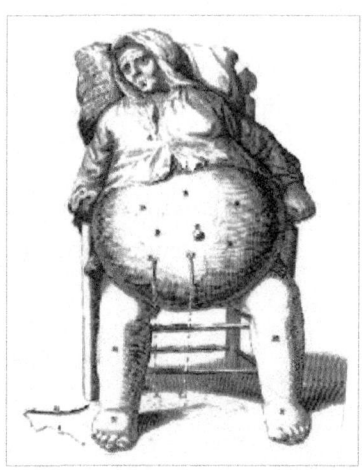

Fig. 2.48 Woman with dropsy whose belly is being "tapped" (paracentesis).
(From Frederik Dekker's [1694] *Exercitationes Practicae Circa Mendendi Methodum*)

four for about 5 minutes to discuss what they see. We come back together and share our observations. Depending on what is brought, the discussion can have different entry points. One observation that I make if it does not come from the class is the odd circumstance that she, as a sick person, is sitting in a chair and not lying in bed. Why might that be? Several reasons can be given, but the primary one is that the mass of fluid extending her stomach would press upward into the lungs if she were lying down and cut off her breath. Not to mention that the lungs themselves might become suffused with water. Because of this, patients were often placed upright in chairs supported by pillows, and they had to live with this situation. Making matters even worse was the fact that every attempt to stoop forward or lean back would give them the feeling of instant suffocation. The inability to move around created an additional problem, one that the students might be able to identify based on the 9th grade biology block. (When we discussed what happens to our muscles—even our bones—if we don't use them.) Her facial expression can be interpreted variously, but the overriding impression is one of, resignation, despondency, maybe even hopelessness. Who could blame her for that?

What is being done for her? She is receiving one of the longest continually used treatments in medical history. Her belly is being "tapped" (paracentesis). Next to her right foot we can see a curved needle that is connected to some kind of string. The needle is surrounded by a sheath and is inserted through the stomach wall and then removed, leaving the sheath behind through which the dropsical fluid can escape.[104] As we observe this woman, we realize how bewildering her situation is: She is not even able to place a basin on the floor beneath her to catch the water flowing out and splashing at her feet[105] (Peitzman 2007).

As our next step, we provide a few drawings of the organ (the kidneys) we are about to discuss, in order to provide the students with a spatial overview of the whole before proceeding to more detailed considerations. First of all, we

---

[104] As background, we find a recommendation found in a medical journal from the late 1300s that says "tappings" should be located "three fingers breadth below the umbilicus." Looking at the picture above, it appears that this advice was being followed.
[105] Although no specific information is available regarding the woman shone above, reports exist of a woman in England with dropsy, who underwent fifty tappings that extracted from her 116 gallons of water.

note that there are two, and they are located to the left and right of the vertebral column, behind the abdomen (retroperitoneal) and below the diaphragm. In the space behind the abdomen we also find the large descending aortic vessel and the inferior vena cava, as well as the vertebral column. Also noticeable is how the right kidney—in deference to the liver—is somewhat lower than the left (about 1.5–2cm.). They rise and fall slightly with each breath. We can see how the large convex outer curvature becomes concave when facing inward (in a gastrula-like gesture), which is an indicator of how strongly the kidneys are oriented toward the large blood vessels passing between them.

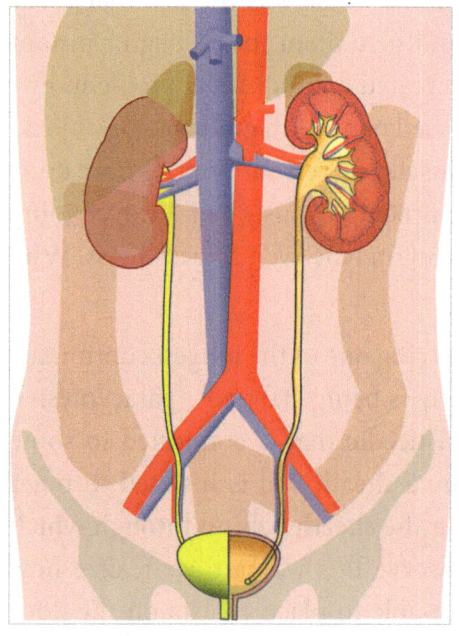

Fig. 2.49 Kidneys oriented toward the descending aorta and the inferior vena cava. (Source: Andrewmeyerson. Creative Commons Attribution-Share Alike 3.0 Unported)

In fact, 20–25% of the blood that passes every moment from the heart into the body flows through the kidneys. The arterial circulation through the kidneys is extraordinary. Located not far from the heart, the renal arteries are the third pair to branch off from the aorta below the diaphragm. Approximately 1500 liters of oxygen-rich blood flows through the kidneys every day. The students already know the pressure-filled, pulsating nature of the blood coming through the aorta. But now, amazingly, we find this tendency amplified even more as the blood flows into the one million glomeruli found in the outer region (cortex) of each kidney.

The blood enters the cup-shaped Bowman's capsules at the periphery of the kidney. Each capsule contains a glomerulus—a cluster of about 50 fenestrated (pore-filled) capillaries. The blood enters through an afferent arteriole that is larger in diameter than the efferent arteriole through which it exits the capsule. This has consequences.

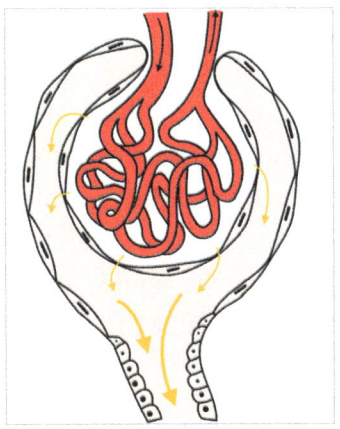

Fig. 2.50 Bowman's capsule with glomerulus and proximal tubule.
(Source: Pharmattila. Creative Commons Attribution-Share Alike 3.0 International)

It creates a blood pressure level in the glomular capillaries that is considerably higher than in capillaries elsewhere in the body. (That an increase in pressure is created by a smaller exit opening can be illustrated by putting your thumb over the end of a garden hose and seeing what happens—something every gardener already knows!)

We reflect together on consequences these conditions could have for the arterial blood. With the teacher's help, we determine that:

1. by entering the small capillaries the pressure increases greatly.
2. Through the dividing up of the afferent vessel into 50-some loops (the glomerulus), an enormous increase in surface area to volume is achieved(!)
3. This brings the blood in contact with capillary walls that are (the teacher tells the students) filled with countless pores.

Put all of this together (the high pressure together with a huge contact area and countless openings) and you get what is called "ultrafiltration." The large pores make the glomular capillaries 100 to 400 times more permeable to the fluid elements of the blood than the capillaries of the skeletal muscles. As a result, under these high-pressure and surface-abundant conditions, water and many solutes—wastes and toxins, as well as substances like glucose and amino acids—pass through the capillary walls (are filtered out) into the capsule. Larger solutes, such as proteins and nutrients, and the blood's cellular components (red blood cells, white blood cells and platelets) are, however, too large to pass through and continue on in the circulating blood.

This is no small matter, since every day we are filtering out of the bloodstream 45–50 gallons (around 2 gallons every hour) of what is called "ultrafiltrate." It is immediately obvious to the students that something more has to happen, and happen fast, since we are filtering out the equivalent of our entire blood plasma volume roughly every 22 minutes. If it were all to pass out of the body—and some of it does—it would mean that we would "urinate ourselves to death" in about 20 minutes! (Tortoro & Dickerson 2013, Van der Graaf & Fox 1998)

So what happens to prevent this? A drawing of an entire nephron helps us follow this process spatially and chronologically.

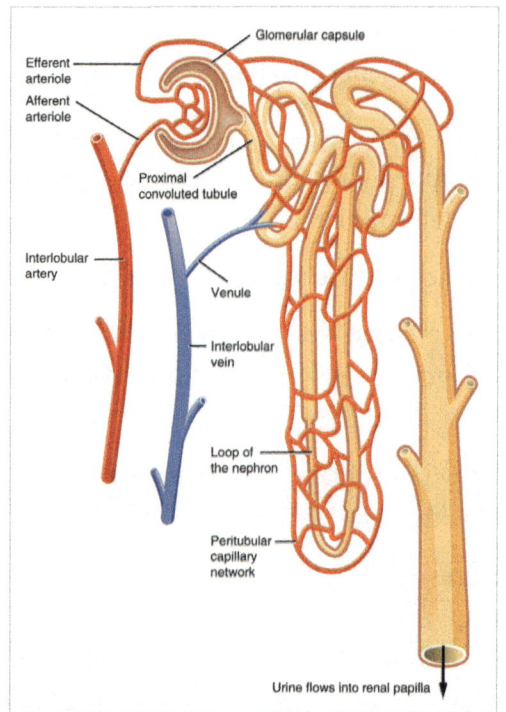

Fig. 2.51 Basic nephron anatomy.
(Source: OpenStax, College Anatomy & Physiology, Connexions Web site. http://cnx.org/content/col11496/1.6/)

As the ultrafiltrate leaves Bowman's capsule, it flows through the Proximal Convoluted Tubule (PCT). Normally the PCT reabsorbs all the amino acids, lactate and glucose in the filtrate, 2/3 of the sodium and water, and the bulk of the electrolytes. They are reabsorbed into the peritubular capillaries that surround the tubules. More water is then reabsorbed on the descending limb of Henle's loop. Further reabsorption of solutes takes place on the ascending limb. By the time the Distal Convoluted Tubule (DCT) is reached, only about 25% of the water remains and 10% of the salt (NaCl).

Most reabsorption from this point on is dependent on the homeostatic needs of the body. If necessary, almost all of the water and salt can be reclaimed as the urine passes down the collecting duct (tubule). This fine-tuning depends on the presence of antidiuretic hormone (ADH), which makes the collecting duct more permeable to water. On the other hand, if blood volume and pressure is high—which is registered by the heart's atrial stretch receptors[106]—ANP (atrial natriuretic peptide) is secreted, which reduces the amount of salt reabsorbed, thereby causing the renal arterioles to absorb less water.

The kidneys are also sensitive to oxygen levels in the blood. As we saw earlier, an increase in the need for oxygen at higher altitudes is met by an increase in the number of RBCs produced. This is possible because the kidneys begin producing the hormone erythropoetin, which stimulates the red bone marrow to produce more RBCs.

Moving down the collecting duct, the urine then passes into the ureters, which bring it to the urinary bladder. The students know what happens from that point on, which means that, suddenly, what is taking place emerges—as if out of nowhere—into human consciousness!

This is quite an amazing process that we have been following. Every day, 45–50 gallons of high pressure blood enters the kidneys, and at the end of the day only about 1.5 liters of urine are eliminated (99% is reabsorbed). With this, many of the body's waste products—in particular the end products of protein metabolism (urea and uric acid)—are excreted.[107] So, instead of "urinating

---

106 That we learned about earlier.
107 In fact, kidney failure is fatal because it leads to uremia, which is the deadly buildup of urea and other nitrogenous waste compounds in the blood.

ourselves to death" as it first appeared we might, our bloodstream has been renewed and brought into balance (homeostasis) through the amazing process of filtration and reabsorbtion that we just learned about. The kidneys, it is now clear, are not only organs of waste removal (excretion), but also of homeostasis, which is essential for the entire organism (Kranich 2003, Marieb & Hoehn 2012, Vogel 1967).

Before we move on to compare the kidneys to another organ we recently learned about, we have an open question still to be resolved. What do the kidneys have to do with the mysterious illness known as dropsy? After all that has just been covered, the students do not have a hard time concluding that the kidneys must be at the root of the problem. It is easy to imagine that the kidneys could err in two directions: reabsorbing too little from the ultrafiltrate or too much (with numerous possibilities in between).

In the case of dropsy (edema), the pivotal factor will be too much reabsorption and too little excretion, of course. There can be many different reasons for this—the heart not sensing accurately an overly high volume of blood entering the right atrium, for example—but the end effect is that the kidneys reabsorb too much salt, and since "water follows salt" (osmotic pressure), more and more water re-enters and overfills the circulating bloodstream. This results in salt and water oozing out into surrounding tissues and the abdominal cavity, creating the kind of swelling (edema) that we observe in dropsy. Case closed (Peitzman 2007) (although we know by now that no issue is ever simply a "nothing but").

Other kidney disorders, for example kidney stones, can be discussed, if time allows. The students are usually interested in hearing about the effects of diuretics like alcohol and caffeine, for example.

Looking Back: Patterns, Polarities, Rhythms

As we near the end of the block, it can be of real value to look back at some of the organs and processes we have covered and see how they appear in relation to one another. We can let them characterize each other, so to speak, and through this we can see if patterns and overriding gestures appear that are symptomatic of the entire organism.

There are many ways to do this. I will begin with one possibility: comparing the last two organs we covered, the kidneys and the liver, and see where it takes us. One of the delightful aspects of intentional comparisons is that they often shed light on qualities and characteristics that have remained in the background until viewed with the new lens created by the comparison.

Looking at the liver with a "kidney lens," one is tempted to say that it has dropsy! It is so plump, so full of fluid, so shapeless when removed from the body. As we saw earlier, its sponge-like structure contains almost as much water (71%) as the blood (78%). What's more, huge quantities of blood are constantly passing through it: about 1.5 liters/minute (~22 gallons/hour). And the fascinating thing is, most of this blood (two-thirds) is oxygen-poor; it is venous blood that is filled with nutrients from the intestinal tract. None of the blood is under high pressure. In fact, it flows much more slowly—7 to 8 times more slowly—through the liver tissue than in normal capillaries, while interacting actively along the way with the liver cells—washing freely around and between them, thanks to the highly porous capillaries. The liver engages and influences the substances passing through it in some 500 different ways, but at the center of all this is its ability to re-synthesizes the nutrients that were broken down in the digestive tract.[108]

If we put this general description side-by-side with the kidneys, several things jump out at us. Although both organs have an enormous amount of blood flowing through them, in the kidneys the blood is totally oxygenated and under very high pressure, which is the opposite of the sleepy flow of oxygen-poor, nutrient-rich blood passing through the liver. In the kidneys the blood

---

108 It is, of course, also an important detoxifier and the synthesizer of bile.

is isolated from all tissue interactions—it has been filtered out—while in the liver, blood to tissue contact is extraordinarily intense. If we look at the organs themselves, a similar contrast is evident. Whereas the liver is a soft, watery "blob," the kidneys are highly structured. At the microscopic level they are, next to the brain, the most highly-structured organ in the body. Whereas the liver has an amazing capacity to regenerate most (80%) of itself,[109] the most the kidneys can produce is scar tissue (Kranich 2003, Marieb & Hoehn 2012, Vogel 1979).

We could go on with this comparison,[110] but in the current context we are looking more for overall gestures/patterns that we have come across during the block. Ones that have not yet been emphasized—or perhaps even noticed. After a swift comparison such as the one above, we can ask the students if it reminds them of other contrasts we have encountered in this block. One common response is that the kidney/liver juxtaposition is similar to the white water/Mississippi river comparison that we used when discussing the difference between blood flow in the arteries and in the veins. The kidneys and arteries share a common dynamic (white water!), while the liver and the veins share a very different, much mellower, Mississippi-like one.

This, in turn, can remind students of the even larger contrast we found between the gravity-free microcirculation in the periphery of the CVS and the muscular, beating heart at its center. We are also reminded of the striking polarity within the heart itself, between the receptive gesture of the venous right heart, and the thick-muscled, pressure-filled, blood-expelling left heart. Dynamically this polarity is also manifest in the systolic and diastolic phases of the heart itself. (It is interesting to do again our full-body rendering of the systole/diastole phases—twisting downward for the systole, unwinding again to open-armed uprightness in the diastole. Students will often sense something of the kidney/liver dynamic in this.)

---

109 The name "liver" comes from *life—to live*. This is no accident, for the name for this organ in many other languages is also a word pertaining to life.

110 For example, we could explore how the liver interacts with carbohydrates, transforming sugar (glucose) to starch (glycogen) and then starch back to sugar again, much like a plant, whereas the kidneys play a major role in regulating nitrogen-based substances, which are central in proteins and characteristic of the animal realm. And so on. Another very interesting comparison for the students can be found in the two organs that are sometimes referred to as "look-alikes": the brain and the small intestine.

The CVS certainly offers us numerous examples of such "polarities." If we step back and ask what we actually mean by the term "polarity," it becomes clear that a polarity is composed of two opposing or contrasting tendencies that work together. We saw already, for example, how the digestive tract breaks down substances, while the liver builds them back up again. The poles need each other in order to exist, in order to do what they do and to play the role they play within the larger whole in which they are embedded—which in our case is the entire human organism. Other classic examples of this are breathing and the beating of the heart: There can be no in-breath without an out-breath, there can be no systole without a diastole. Take the one pole by itself and you have death.

It is helpful to remind the students at this juncture of what we discussed earlier regarding the left heart, that during the systolic phase its inner layers only receive full circulation (perfusion) 20% of the time. This very one-sided, almost deadly condition is, however, balanced out by the total relaxation that takes place during the diastolic phase. From there we spoke about a number of "larger" polarities (rhythms) that provide a context for most of our (and the natural world's) existence: summer↔winter, day↔night, sleeping↔waking, vacation↔school, boring classes↔biology class, etc. We also found this in our inner life. We noticed how our feelings always have an opposite pole: happy↔sad, interest↔boredom, sympathy↔antipathy, laughing↔crying, etc.

Once we "wake-up" to the pattern of polarities that move rhythmically between one-sided extremes, we can begin to see how they manifest not only within an individual organ, or organ system, but at many different levels.

We can even look back to our 9th grade block and recognize how in our limbs we have a polarity between the structural pole found in the bones and movement pole in the muscles—mediated by the joints. Or we think back to how the head-bearing vertebrae (atlas and axis) are small and mobile, while at the load-carrying pole—the lumbar vertebrae—the vertebrae are massive and stabile. And, of course, we see a classic polarity between the spherical—brain-enclosing—head, with only one mobile joint (the jaw), and the radial, highly dynamic limbs that carry the head around. While the head pole takes in the world through the senses and reflects upon it, the limbs take us out into the world and engage in it. The intentions we arrive at with the help of our head pole, are brought to fruition with the help of our limbs, and through the experiences made there, we are able to correct or confirm our thinking. This influences the way we act next time around. Even though the two poles interact in totally different ways with the world around us, they constantly educate and learn from each other.

Interestingly, in the limb pole itself we can also find two poles that work together: the stabilizing, gravity-embedded lower limbs (legs and feet) at one extreme, the dynamic, for the most part gravity-emancipated upper limbs (arms & hands), at the other. Which is to say, within primary polarities (such as head←→limbs), we often find secondary polarities, such as upper limbs←→lower limbs, arms←→hands, cranium←→jaw, and so on.

The activity of the muscles also shows this. On the one hand, every muscle moves between two extremes, between contraction and relaxation. But in addition to that, around each joint you must have two muscles working in a polar relationship to each other—in the arms, for example, the biceps and triceps. When the one contracts, the other must relax, and vice versa. And so we could go on and on with many different examples at many different levels.[111]

The essential thing in this context—at the end of our human biology block sequence in 9th and 10th grades—is to bring what were rather isolated considerations of different aspects of our anatomy and physiology back together again by viewing the whole from a dynamic perspective such as polarity. Through such an overriding principle—that manifests in countless ways—it is possible to bring life not only into the student's understanding of an organism, but also into the way they see nature as a whole. In this way, dynamic interactions are highlighted rather than a catalog of anatomical facts.

When discussing polarities, one final aspect needs to be emphasized. Although implied in much of what has just been described, it is worth emphasizing more explicitly the "time dimension" that unites the two extremes

---

[111] In the 10th grade earth science block, there are also many polarities to be observed that shape the earth's biosphere. Beyond the obvious polarity of high and low pressure systems in the atmosphere, one particularly fascinating polarity is found in the contrast between the rainforest belt at the equator and the deserts found at approximately 30 degrees north and south of the equator. This polarity is, as we know, an expression of the Hadley Cell of the Intertropical Convergence Zone (ITCZ).

of a polarity. Through this temporal relationship—known as "rhythm"—the poles are not just two isolated extremes in a range of possibilities, but are engaged in a living dynamic that swings between them. Although the rhythm never repeats itself exactly, the two poles are interconnected by the extent to which each one manifests. A pendulum can provide us with a clear picture of this: the degree to which it swings out to the left is directly related to the distance it swings to the right, and vice versa. We see such a relationship in an organic context in the way that the depth of an in-breath is directly related to the extent of the out-breath that preceded it (unless a conscious attempt is made to intervene in the breathing process). We can find many variations on this theme, the relationship of the crest of a wave to its trough, the relationship of the need for sleep to the time one was awake (and to the quality of that wakefulness).[112] When the two poles fall out of harmony with each other, this usually manifests as illness.[113] That climate change might have to do with such a "falling out of harmony" will seem quite evident to many of the students. (How can rainforests be destroyed at the rate they have been for decades without throwing something out of balance—which then throws something else out of balance, and so on?)

---

112 There are also rhythms that alternate between the gradual dominance (emphasis, presence) of one pole and then the gradual dominance of the other. Day and night, for example, do this in the course of the year—and with them summer and winter.

113 This is a huge and very important topic, but one that cannot be explored in depth in this block. The Austrian biologist, Otto J. Hartmann, captures this perspective in a few words:

"In the living human being we not only have different forces at work, whose emphasis is in different organs and functions, but these work in ways that are the polar opposite of each other. Our entire physiology is full of such polar opposites. For every organ and process in the human being, it is possible to find a corresponding one that is the opposite of it....

"This is the context in which the relationship of illness and health must be understood. Life and health are not stable conditions, but rather highly sensitive states of equilibrium that must be achieved again and again—and in the process of achieving such balance, it will be lost once more. The root of health and illness ... lies in the interpenetration of opposing processes and bodily constituents. Humans grow ill primarily through disturbances of their inner equilibrium. Such equilibrium is rhythmical, and therefore the rhythmical organ systems (heart & circulation, breathing, also waking-sleeping) are decisive for our health. By contrast, all arrhythmia leads to illness" (Hartmann 1959, pp. 38–39; transl. by MH).

114 In fact, Gunther Hildebrandt, the founder of the International Society for Chronobiology and Professor of Medicine at the University of Marburg in Germany, discovered through decades of research that rhythmical processes in the human being are themselves rhythmically organized. He found that a polarity exists between the rapid and highly variable rhythmical oscillations of the Central Nervous System centered in the head and the much slower, much more stable rhythms of the metabolic organs centered in the lower half of the thorax. These two extremes are mediated by the heart and lung rhythms that lie between the two poles not only spatially, but in their frequency and stability (Hildebrandt 1986, Schad 2014).

It is not difficult to come to the conclusion that organisms are systems of complex interacting rhythms—a perspective that is focused upon in the growing field of Chronobiology.[114] It relates to something that has come up again and again in this block: the importance of keeping an eye on the whole organism (the forest), while learning about many details (the trees) along the way. I think emphasizing polarities and rhythms is one way to help meet this challenge. Living organisms are verbs, not nouns. They are lawfully organized processes that manifest in time through the rhythmical balancing/harmonizing of a multitude of functions (Rosslenbroich 1994).

As one of the pioneers of 20th century phenomena-based biology, Jochen Bockemühl, put it:

Any time we have to do with living nature, rhythms come to meet us. ... Rhythm is "the element" of life itself. Without it, life cannot exist. All life comes forth out of rhythms. For that reason, we, too, come closer to the life of organisms, when we study their rhythms. (1982, p. 26)

Coming to a close, I would like to quote a few passages from the 9th lecture of Rudolf Steiner's *Study of Man* that are particularly pertinent for biology teachers. Particularly pertinent in a time when many people carry around with them "finished concepts" that are of the "nothing but" genre (the heart is "nothing but" a pump) referred to at the beginning of this book. These are the kind of "dead concepts" that close people off from the complex realities of life. Instead, Rudolf Steiner tries to awaken our awareness for the significance of another kind of understanding:

The educator must aim at giving young people concepts which will not remain the same throughout their lives, but will evolve as they grow older. If you do this you will be implanting living concepts in the children. And when is it that you give them dead concepts? When you continually give the children definitions, when you say: "A lion is ..." this or that, and make them learn it by heart, then you are inoculating dead concepts into them; and you are expecting that at the age of thirty they will retain these concepts in the precise form in which you are now teaching them. The making of many definitions is death to living teaching. What then must we do? In teaching we must not make definitions but rather must attempt to make characterizations. We characterize things when we view them from as many standpoints as possible.

In a reasonable curriculum this characterization will arise of itself, if, for instance, the teacher does not merely describe consecutively, say: first the cuttlefish, and then the mouse, and finally the human, each in turn, in natural-historical order—but rather places cuttlefish, mouse and human being side-by-side and relates them one with the other. The interrelationships will prove so manifold that there will result, not a definition, but a characterization. A right kind of teaching will aim, from the outset, at characterization rather than definition.

It is of very great importance to make it your constant and conscious aim not to destroy anything in growing human beings, but to teach and educate them in such a way that they continue to be full of life, and do not dry up and become hard and inflexible… to equip them with living concepts that are open to change, that undergo metamorphosis, that transform themselves throughout a person's entire life. (Steiner 1990, pp. 131–133; slight modifications made to translation by MH)

Bibliography – Part II

Akst, J. (2014). Microbes of the Skin. *The Scientist*. June 13.

Albonico, H. and Lemann, D. (1993). HIV-AIDS. *JAM*. Vol. 10 Nr. 2.

Aschoff, J. (1965). Circadian Rhythms in Man. *Science* 148: 1427–1432.

Benninghoff, A. and Goerttler, K. (1980). *Lehrbuch der Anatomie des Menschen, II. Band*. München: Urban & Schwarzenberg.

Berra, Y. (1974). T*he Wit and Wisdom of Yogi Berra*. Westport, CT: Mecler Books.

Biga, Dawson, et al. (2019). *Anatomy & Physiology*. Creative Commons Attribution-ShareAlike 4.0 International License.

Blaser, M. (2014). *Missing Microbes: How the Overuse of Antibodies is Fueling Our Modern Plagues*. New York: Henry Holt & Co.

Brettschneider, H. (2002). The Polarity of Center and Periphery in the Circulatory Stystem. In Brown, Smallwood, et al. (1999). *Medical Physics and Biomedical Engineering*. New York: Taylor & Frances.

Bockemuehl, J. (1982). Lebensrhythmen bei Pflanze und Tier. In W. Schad (ed.) *Allgemeine Biologie*. Stuttgart: Freies Geistesleben.

Bos, A. (1989). *AIDS*. Stroud, UK: Hawthorn Press.

Bott, V. (2004). *Anthroposophical Medicine*. Forest Row, GB: Sophia Books.

Cacioppo, Berntson, et al. (2000). *The Psychophysiology of Emotion*. www.researchgate.net.

Chopra, D. and Tansi, R. (2015). *Super Genes*. New York: Crown Publishing Group.

Costanzo, L. (2013). *Physiology*. Philadelphia: Saunders.

Davis, N. (2020). Severe Childhood Deprivation Reduces Brain Size. *The Guardian*. 6 Jan. 2020.

DeSalle, R. and Perkins, S. (2015). *Welcome to the Microbiome*. New Haven: Yale University Press.

Dietert, R. (2016). *The Human Superorganism*. New York: Dutton.

Doidge, N. (2007). *The Brain that Changes Itself*. New York: Viking.

Douglas, A. (2018). *Fundamentals of Microbiome Science*. Princeton: Princeton University Press.

Eckert, D. and Butler, J. (2017). *Principles and Practice of Sleep Medicine*. Amsterdam: Elsevier.

Ferranti, E. et al. (2014). "20 things you didn't know about the human gut microbiome." *The Journal of Cardiovascular Nursing,* vol. 29,6 (2014): 479–481.

Ferrell, J. and Chiang J. (2015). Circadian rhythms in liver metabolism and disease. *Acia Pharm Sin B*. 5(2): 113–122.

Fields, R. (2020). "The Brain Learns in Unexpected Ways" in *Scientific American* 322, 3, 74–79 (March 2020).

Fonken, L. and Nelson, R. (2014). The effects of light at night on circadian clocks and metabolism. *Endocr Rev*. 35: 648–670.

Furst, B. (2014, 2020). *The Heart and Circulation—An Integrative Model*. Switzerland: Springer.

George, R. (2020). *Nine Pints*. London: Granta Books.

Grice, A.; Kong, H.; Conlan, S. et al. (2009). Topographical and Temporal Diversity of the Human Skin Microbiome. *Science*. 2009 May 29: 324(5931): 1190–1192.

Grice, E. and Segre, J. (2011). The Skin Microbiome. *Nat Rev Microbiol*. 9(4): 244–253.

Hartmann, O. (1959). *Menschenkunde*. Frankfurt: Vittorio Klorsterman.

Harvard School of Public Health (2020). *The Microbiome*. https://www.hsph.harvard.edu.

Hickman, C.; Roberts, L. and Larson, A. (2001). *Integrated Principles of Zoology*. New York: McGraw-Hill.

Hildebrandt, G. (1986). Zur Physiologie des rhythmischen Systems. *Beitraege zu einer Erweiterung des Heilkunst* 39, 8–29.

Hirsch, R. (2017). *The Human Microbiome*. Minneapolis: Lerner Publications.

Holdrege, C. (ed.) (2002). *The Dynamic Heart and Circulation*. Fair Oaks, CA: AWSNA Publications.

Holdrege, M. (2014). "A New 'Land Ethic' for Teaching Embryology in Grade 10." In D. Gerwin (ed.), *Trailing Clouds of Glory*. Chatham, NY: Waldorf Publications.

Husemann, F. and Wolff, O. (1987). *The Anthroposophical Approach to Medicine*. Vol. II. Hudson, NY: Anthroposophic Press.

Jensen, F. (2015). *The Teenage Brain*. New York: Harper Collins.

Kandel, E.; Schwartz, J. and Jessel T, (2000). Principles of Neural Science. New York: McGraw Hill.

Kettner et al. (2016). Circadian Homeostasis of Liver Metabolism Suppresses Hepatocarcinogenesis. *Cancer Cell* 30, 909–924. Elsevier, Inc.

Keysers, C. (2011). *The Emphatic Brain*. Social Brain Press.

Klobunde, R. (2017). *Coronary Anatomy and Blood Flow*. https://www.cvphysiology.com/Blood%20Flow/BF001

Korade, Z. and Mirniks, K. (2014). Programmed to be Human? *Neuron*, Vol. 82 (2), January 22, 2014.

Kranich, E. (2003). *Der innere Mensch und sein Leib*. Stuttgart: Verlag Freies Geistesleben.

Kulkarni, S.; O'Farrell, I.; Erasim, M. and Kochar, M. (1998). Stress and hypertension. *WMJ*. 1998; 97(11): 34–38.

Lauboeck. H. (2002). The Physiology of Circulation. A Reappraisal. In C. Holdrege (ed.), *The Dynamic Heart and Circulation*. Fair Oaks, CA: AWSNA Publications.

Leisman, G. and Melillo, R. (2015). Infant and Childhood Frontal Lobe Development: Asymmetry and the Regulation of Temperament and Affect. *Functional Neurology Rehabilitation and Ergonomics*. Vol 5.

Le Doux, J. (2002). *Synaptic Self: How Our Brains Become Who We Are*. New York: Penguin. Depression, Stress, Immunity (1987). *The Lancet*, Vol. 992, June 27.

Lazaroff, M. (2004). *Anatomy and Physiology*. New York: Alpha.

Llewellyn, K. (1994). *Atlas of the Human Body*. New York: Harper Collins.

Manteuffel-Szoege, L. (1977). *Über die Bewegung des Blutes*. Stuttgart: Verlag Freies Geistesleben.

Marieb, E. and Hoehn, K. (2012). *Human Anatomy and Physiology*. San Francisco: Pearson.

McCraty, R. (2015). *Science of the Heart*, Vol. 2. Publisher: Heartmath.

McGilchrist, I. (2009). *The Master and His Emissary*. New Haven: Yale Univ. Press.

Meyer, K. (2015). In der DDR gab es nur halb so viel Allergien wie im Westen. *Badische-Zeitung*.de. 10/3/15.

Moeller, K.; Willmes, K. and Klein, E. (2015). A review on functional and structural brain connectivity in numerical cognition. *Front. Hum. Neurosci.* 9:227.

Moerike, K.; Betz, E. and Mergenthaler, W. (2007). *Biologie des Menschen*. Verlag Nikol.

Moise, A. (2017). *The Gut Microbiome*. Santa Barbara, CA: Greenwood.

Mosby's *Medical Dictionary* (2006). St. Louis: Elsevier.

Norris, M. and Sigfried, D. (2011). *Anatomy and Physiology*. Hoboken, NJ: Wiley & Sons.

Pasipoularides, A. (2010). Heart's Vortex: Intracardiac Blood Flow Phenomena, 301–302. Cited in B. Furst. (2020).

Peitzman, S. (2007). *Dropsy, Dialysis, Transplant*. Baltimore: Johns Hopkins Univ. Press.

Perry, B. (2002). Childhood Experience and the Expression of Genetic Potential: What Childhood Neglect Tells Us About Nature and Nurture. *Brain and Mind*, Vol. 3, 79–100.

Pollan, M. (2013). Some of My Best Friends are Germs. *The New York Times Magazine*. May 15, 2013.

Ratey, J. (2002). *A User's Guide to the Brain*. New York: Random House.

_____. (2008) *Spark – The Revolutionary New Science of Exercise and the Brain*. New York: Little, Brown & Co.

Restak, R. (1995). *Brainscapes*. New York: Hyperion.

_____. (2006). *The Naked Brain*. New York: Random House.

_____. (2009). *Think Smart*. New York: Riverhead Books.

_____. (2010). *The Playful Brain*. New York: Riverhead Books.

Rohen, J. (2007). *Functional Morphology*. Hillsdale, NY: Adonis Press.

Rosslenbroich, B. (1994). *Die rhythmische Organization des Menschen*. Stuttgart: Freies Geistesleben.

Ruskin, J. (2012). The Storm Cloud of the Nineteenth Century. Pallas Athene: Reprint edition. Cited in E. Lehrs (1958). *Man or Matter*. London: Faber & Faber.

Sadler, T. (2000). *Langman's Medical Embryology*. Baltimore: Lippencott, Williams & Wilkins.

Schad, W. (2014). *Der Periphere Blick*. Stuttgart: Freies Geistesleben.

Schaefer, K. (1979). Individual Respiratory Patterns Affecting Metabolic Processes and CNS Functions. In Schaefer, Hildebrandt, & Macbeth (eds.), *Basis of an Individual Physiology*. Mt. Kisko, NY: Futura Publishing.

Schmidt, G. (1975). *Dynamische Ernährungslehre*. St. Gallen: Proteus Verlag.

Schoeffler, H. (1975). *Die Zeitgestalt des Herzens*. Stuttgart: Verlag Freies Geistesleben.

Schoorel, E. (2004). *The First Seven Years. Physiology of Childhood*. Fair Oaks, CA: Rudolf Steiner College Press.

Schwarz, J. and Begley, S. (2003). *The Mind and the Brain: Neuroplasticity and the Power of Mental Force*. Harper Perennial.

Sender, R.; Fuchs, S. and Milo, R. (2016). Revised Estimates for the Number of the Body. Human and Bacteria Cells in *t PLoS Biol*. 2016 Aug. 14(8).

Shapira, M. (2016). Gut Microbiota and Host Evolution: Scaling up Symbiosis. *Trends in Ecology and Evolution*. 31 (7): 539–549.

Steiner, R. (1919). *The Study of Man*. London: Rudolf Steiner Press. (CW 293)

_____. (1921). *Faculty Meetings with Rudolf Steiner*. Hudson, NY: SteinerBooks. (CW 300b)

_____. (1921). *The Impulse for Renewal in Culture and Science*. Rudolf Steiner Publications. (CW 79)

_____. (1996) *Education for Adolescents*, CW 302. Great Barrington, MA: Anthropsophic Press.

Stromberg, J. (2014). Five Reasons Why You Should Probably Stop Using Antibacterial Soap. *Smithsonianmag.com*, Jan 3.

Spruill, T. (2010). Chronic psychosocial stress and hypertension. *Current hypertension reports*, 12(1), 1 https://doi.org/10.1007/s11900-16. 6-009-0084-8

Tyng, C.; Amin, H.; Saad, M. and Malik, A. (2017). *The Influences of Emotion on Learning and Memory*. Front. Psychol. 8:1454.

Totoro, G. and Derrickson, B. (2013). *Principles of Anatomy and Physiology*. Hoboken, NJ: John Wiley & Sons.

Van der Graaf, K. and Fox, S. (1998). *Concepts of Human Anatomy and Physiology*. Chicago: Wm. C. Brown Publ.

Vogel, L. (1979). *Der Dreigliedrige Mensch*. Dornach: Philosophisch-Anthroposophischer Verlag.

Von Mutius, E. (2007). Asthma and Allergies in Rural Areas of Europe. *Proc Am Thorac Soc*. Jul 4 (3): 212–216.

Von Mutius, E. and Vercelli, D. (2019). Farm Living: Effects on Childhood, Asthma and Allergy. *D. Nat Rev Immunol*. Sep 19 (9).

Wolff, O. (1987). The Cardiovascular System. In Husemann & Wolff (eds.), *The Anthroposophical Approach to Medicine*. Vol. II. Hudson, NY: Anthroposophic Press.

_____. (1993). The Immune System and Inner Activity. *JAM*. Vol. 10 Nr.2.

Yong, E. (2010). An Introduction to the Microbiome. *Natl. Geogr*. Aug. 6, 2010.

Index

20th century view of the brain 190, 198

A

accommodation 26
achieving uprightness 78-82
ambiguous figures 33
ANP (atrial natriuretic peptide) 233
antibiotic use, excessive 218-221
arborization 191
arteries and veins 121
atmospheric (structural) colors 24
auditory cortex 43

B

bacterial "starter set" 217
baroreceptors 142
bats 37
bile 227
biological self (individuality) 172, 175
Blaser, Martin 218
blood
    central functions 157-158
    components 156-157, 158-165
    interfacing, mediating & connecting 126
    pressure 154-156
blushing 148
Bockemühl, Jochen 240
Bohm, David 105
Bos, Arie 176
Bowman's capsule 231
brain 186-202

C

calling on inner will activity 121, 128
capillaries 124-126
capillary density in major organs 151
cardiac output & pacemaker tachycardia 137
cardiovascular system (CVS) 116-165
CD4 count 175-176
Ceaucescu, Nicolai 197

changing blood pressure with age 155
characterizations rather than definitions 240-241
chemoreceptors 142
chicken or egg? 116
childhood: the immune system goes to school 172, 217
chronobiology 240
circulation 118
comparing and contrasting 22, 66, 67-77, 235-236, 241
complex, multifaceted judgments 112, 113
concepts that will evolve as students grow older 240
coronary circulation 152-154
COVID-19 178
Crick, Frances 105-106

D

describing not naming 73
Dietert, Rodney 219
different levels of muscular activity 92-98
digestive system 202
downside-up, outside-in tree 180
dropsy 228
dynamic interactions vs. cataloged facts 238
dynamic, fluid thinking 116

E

ear and hearing 36-46
ecological approach 213
embryological formation of the heart 131
embryonic brain development 190-191
emergence 108
emerging capacities 108
engaging the student's feelings 162-163, 169, 213-214
equilibrium 47, 52, 240

Index

equip children with living concepts that are open to change 241
exchange surface for microcirculation 125
exemplary learning 11-13, 153
eye and seeing 22-25

F

facial expressions 96
factory farming 107
farsightedness 28
feelings and heart rate 147-148
fever 149-150, 166
figuration 18, 33, 60
fluids secreted into digestive tract 210
Frankel, Viktor 107, 112
Furst, Branko 116, 132, 140

G

gestures 96
Giedd, Jay 199
glucose and glycogen 224-225
Goethe's Theory of Color 23
gravitation 48, 51, 64

H

Hartmann, O.J. 239
head pole – skeleton 72-75
heart 126-135
heart as used in the English language 148-149
Hildebrandt, Gunther 239
HIV/AIDS 172, 174-178
human immune system 166-178
Human Microbiome Project (HMP) 213
human skeleton 64-82
Husemann & Wolff 135, 137
hypothermia 149
hypothetical-deductive thinking 9, 16

I

idealism 10
illness and health 239
inertia 49-55, 58

inner and outer 147
inner experiences 113, 173
inner organs 20-21, 114-115
inner-life and breathing 183-185
insulin 225
interhemispheric communication 200
interneurons (association fibers) 193-194

J

Jenner, Edward 168
Jensen, Frances 199
joints 83-87
judgment-formation 9-13

K

kidneys 228-234
Kranich, E.M. 93

L

large intestine 212
largest gland in the body 222
larynx and speech 59-63
larynx and tongue 97-98
less is more 11-13, 153
leverage 17, 40-41, 88-92
limb pole – skeleton 67-72
liver 222-227
    capacity to regenerate 222
    gatekeeper 223
    24-hour circadian rhythm 225-227
long-term potentiation (LTP) 196
looking at organs & processes in relation to each other 235-236

M

Manteuffel-Szoege, Leon 117
mechanical advantage & disadvantage 92
microbiome 213-221
microcirculation 124, 139
    and flowback 140-142
mind-body problem 115
mobile thinking 113
modern plagues 218

Index

Mount St. Helens 141
mouth and esophagus 202
muscles working together in flowing unity, like schools of fish 92, 95
muscular system 88-98
myelination 197
myocardial infarction 154

N

nearsightedness 28
nephron 232
neural plasticity 194-197, 200
non-specific (innate) immune system 164, 166-168, 170
nothing-but-ness 107, 108, 112-116, 174, 234, 240

O

one-dimensional judgments 116
ontological discontinuities 108
organ of human substance 224
organisms
    as systems of complex interacting rhythms 240
    as verbs, not nouns 240
organizing activity of the mind 33-34
organs as instruments 115
outer organs 20-21, 114-115
overall patterns and gestures 236
oxygen consumption correlates with cardiac output 138-139

P

paracentesis (belly tapping) 229
Pasipoularides, Ares 132
Pasteur, Louis 168
path of the blood flow 120-121, 129
patterns, polarities, rhythms 235-241
peripheral circulation 136-140
peristalsis 205
phenomenology 10, 11
phonation and articulation 63

polarities
    and rhythms 143-46
    primary and secondary 238
polarity characterized 237
Pollan, Michael 219
practice, practice, practice 54, 95, 195-197
pre-frontal cortex 199
Pressel, Simon 92
pressure wave distinguished from blood flow 136
process-based understanding 112
proprioception 53-58, 59
pruning 195
psychoneuroimmunology 174

Q

quick glimpse method 72-74

R

Ratey, John 186
reductionism 104-108
respiratory system 179-185
Restak, Richard 200
rhythms 135-137, 237
riddles 16
Rohen, Johannes 63

S

saccadic eye movements 57
Sachs, Oliver 52, 54
Schumacher, E.F. 107
second wave of synaptic sprouting 198
self-absorption 97
sense of balance (equilibrium) 47-52
sense of movement (proprioception) 53-58, 59
sense perception 18
senses 17, 18, 22
sensing form 57-58
sensitive states of equilibrium 240
sharing air 179
skeletal/muscular system 16, 17

skeleton 64-82
    overview 64-67
skin: a complex ecosystem 166
Sloan, Douglas 106-107
small intestine 208
sound wave frequency, pitch 37, 41-45
specific (adaptive) immune system 164, 168-178
speech centers of the brain 63
spiraling, vortex-creating flow patterns 129-133
starting with the whole 118
Steiner, Rudolf 97, 112, 113, 162-163, 240-241
stomach 206
straightforward causality 18, 26, 28, 39, 45-46, 51, 60, 64, 112, 113
Strauss, Erwin 82
surface area 41
    of intestinal folds 211
    to volume (SA:V) 125, 140
synaptogenesis 191-192, 195

T

teens determine their own brain development 199
thinking 18
time dimension 238-239
torso – skeleton 75-78

U

ultrafiltration 231
uprightness 49, 68, 78-82

V

vaccinations 173
vasodilation 139
venous heart of a fish 118
vertebral column 75-82
visual cortex 34-35
von Mutius, Erika 216-217

W

Wagenschein, Martin 11-13, 153
warmth creation in major organs 150
warmth organization 149-152
white water 121
Whitehead, A.N. 105
Wolff, Otto 136

Acknowledgments

In addition to my teachers mentioned in the preface, I am also indebted to two lifelong friends and fellow explorers in a wide range of questions—Craig Holdrege and Jeff Martin. Craig and Jeff were not only willing to read through the manuscript, but also offered valuable editing suggestions for its betterment. At Waldorf Publications, many thanks go to Patrice Maynard, who guided this project with a steady and thoughtful hand, and to Ann Erwin for her patience and skill in solving numerous layout issues. Lastly, I wish to thank my wife Patricia for her ongoing equanimity when confronted with a living room suffering from perpetually shifting stacks of books, ones that eventually began competing for spaces intended for her lovely plants.

Made in the USA
Monee, IL
22 March 2023

29638862R00142